U0362882

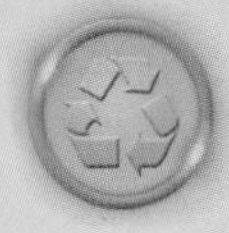

Research Series of Key Technologies on
Energy Saving and New Energy Vehicles
Editor-in-Chief: Minggao Ouyang

Jian Chen
Quan Ouyang
Zhisheng Wang

Equalization Control for Lithium-Ion Batteries

華中科技大學出版社
http://press.hust.edu.cn
中国 · 武汉

Springer

图书在版编目(CIP)数据

锂离子电池均衡控制=Equalization Control for Lithium-Ion Batteries / 陈剑,欧阳权,王志胜著. -- 武汉 : 华中科技大学出版社, 2024. 6. --(节能与新能源汽车关键技术研究丛书).
ISBN 978-7-5772-0941-8

Ⅰ. TM912

中国国家版本馆 CIP 数据核字第 202448ZD97 号

锂离子电池均衡控制
LILIZI DIANCHI JUNHENG KONGZHI

陈 剑 欧阳权 王志胜 著

策划编辑:俞道凯 胡周昊
责任编辑:杜筱娜
责任监印:朱 玢
出版发行:华中科技大学出版社(中国·武汉) 电话:(027)81321913
武汉市东湖新技术开发区华工科技园 邮编:430223
录 排:武汉三月禾文化传播有限公司
印 刷:武汉科源印刷设计有限公司
开 本:710mm×1000mm 1/16
印 张:13
字 数:328 千字
版 次:2024 年 6 月第 1 版第 1 次印刷
定 价:158.00 元

華中科技大學出版社
Huazhong University of Science and Technology Press

Website: http://press. hust. edu. cn
Book Title: Equalization Control for Lithium-Ion Batteries

Contact address: No. 6 Huagongyuan Rd, Huagong Tech Park, Donghu High-tech Development Zone, Wuhan City 430223, Hubei Province, P. R. China.
Phone/fax: 8627-81339688 **E-mail**: service@hustp. com

First published: 2024
ISBN: 978-7-5772-0941-8

Cataloguing in publication data: A catalogue record for this book is available from the CIP-Database China.

Printed in the People's Republic of China

Committee of Reviewing Editors

Foreword: New Energy Vehicles and New Energy Revolution

The past two decades have witnessed the research and development (R&D) and the industrialization of China's new energy vehicles. Reviewing the development of new energy vehicles in China, we can find that the Tenth Five-year Plan period is the period when China's new energy vehicles began to develop and our nation started to conduct organized R&D of electric vehicle technologies on a large scale; the Eleventh Five-year Plan period is the stage that China's new energy vehicles shifted from basic development to demonstration and examination as the Ministry of Science and Technology carried out the key project themed at "energy saving and new energy vehicles"; the period of the Twelfth Five-year Plan is the duration when China's new energy vehicles transitioned from demonstration and examination to the launch of industrialization as the Ministry of Science and Technology organized the key project of "electric vehicles"; the period of the Thirteenth Five-year Plan is the stage when China's new energy vehicle industry realized the rapid development and upgrading as the Ministry of Science and Development introduced the layout of the key technological project concerning "new energy vehicles".

The decade between 2009 and 2018 witnessed the development of China's new-energy automobile industry starting from scratch. The annual output of new-energy vehicles developed from zero to 1.27 million while the holding volume increased from zero to 2.61 million, each of which occupied over 53% in the global market and ranked 1st worldwide; the energy density of lithium-ion power batteries had more than doubled and the cost reduced by over 80%. In 2018, 6 Chinese battery companies were among the top 10 global battery businesses, with the 1st and the 3rd as China's CATL and BYD. In the meanwhile, a number of multinational automobile businesses shifted to develop new-energy vehicles. This was the first time for China to succeed in developing high-technology bulk commodities for civic use on a large scale in the world, also leading the trend of the global automobile development. The year of 2020 marked the landmark in the evolution of new-energy automobile. Besides, this year was the first year when new energy vehicles entered families on a large scale and the watershed where the new-energy vehicle industry shifted from policy-driven to market-driven development. This year also saw the successful wrapping up of the mission in the *Development Plan on Energy Saving and New Energy Vehicle Industry (2012–2020)* and the official release of *Development Plan on New Energy Vehicle Industry (2021–2035)*. At the end of 2020, in particular, president Xi Jinping proposed that China strove to achieve the goal typified by peak carbon dioxide emissions by 2030 and carbon neutral by 2060, so as to inject great power into the sustainable development of the new energy vehicle industry.

Looking back to the past and looking forward to the future, we can see even more clearly the historical position of the current development of new energy vehicles in the energy and industrial revolution. As is known to us all, each and every energy revolution started from the invention of power installations and transportation vehicles. On the other hand, the progress of power installations and transportation vehicles contributed to the development and exploitation of energy and led to industrial revolutions. In the first energy revolution, steam engine was used as the power installation, with coal as energy and train as the transportation. As for the second energy revolution, internal combustion engine was taken as the power installation, oil and natural gas as energy, gasoline and diesel as energy carriers, and automobile as the transportation vehicle. At the current stage of the third energy revolution, all kinds of batteries are power installation, the renewable energy as the subject of energy and electricity and hydrogen as energy carriers, and electric vehicles as the means of transportation. In fact, the first energy revolution enabled the UK to outperform Netherland while the second energy revolution made the USA overtake the UK, both were in terms of the economic strength. The present energy revolution may be the opportunity for China to catch up with and surpass other nations. How about the fourth industrial revolution? In my opinion, it is the green revolution based on renewable energy and also the smart revolution on the basis of digital network.

From the perspective of energy and industrial revolution, we can find three revolutions closely related to new energy vehicles: electrification of power—the revolution of electric vehicles; low-carbon energy—the revolution of new energy; systematic intelligence—the revolution of artificial intelligence (AI).

Firstly, electrification of power and the revolution of electric vehicles.

The invention of lithium-ion battery triggered the technological revolution in the area of storage battery over the past 100 years. Viewed from the development of power battery and power electronic device, the involvement of high specific energy battery and high specific power electric drive system would contribute to the platform development of electric chassis. The volume power of the machine controller based on new-generation power electric technology has more than doubled to 50 kW. In future, the volume power of the high-speed and high-voltage machine can be nearly doubled to 20 kW and the power volume of the automobile with 100 kW volume power could be no more than 10 L. With the constant decline of the volume of the electric power system, the electrification will lead to the platform and module development of chassis, which will lead to a major change in terms of vehicle design. The platform development of electric chassis and the lightweight of body materials will bring about the diversification and personalization of types of vehicles. Besides, the combination of active collision avoidance technology and body lightweight technology will result in a significant change in automobile manufacturing system. The revolution of power electrification will promote the popularity of new energy electric vehicles, and will eventually contribute to the overall electrification of the transportation sector. China Society of Automobile Engineers proposed the development goals of China's new energy vehicles in the *2.0 Technology Road Map of Energy Saving and New Energy Vehicles*: the sales of new energy vehicles would reach 40% of the total sale of vehicles by 2030; new energy vehicles would become the mainstream by 2035 with its sale accounting for over 50% of the total sale of vehicles. In the foreseeable future, electric locomotives, electric ships, electric planes and other types will become a reality.

Secondly, low-carbon energy and the revolution of new energy.

In the Strategy on Energy Production and Consumption Revolution (2016—2030) jointly issued by National Development and Reform Commission and National Energy Administration, a target was proposed that the non-fossil energy would account for around 20 percent of total energy consumption by 2030 and over 50% by 2050. Actually, there are five pillars aimed to realize the energy revolution: first, the transition from traditional resources to renewable resources and the development of photovoltaic and wind power technologies; second, the transformation of energy systems from centralized to distributed development which can turn every building into a micro-power plant; third, the storage of intermittent energy by using of technologies related to hydrogen, battery, etc.; fourth, the development of energy (electric power) Internet technology; fifth, enabling electric vehicles to become the end of energy usage, energy storage and energy feedback. In fact, China's photovoltaic and wind power technologies are fully qualified for large-scale distribution, but energy storage remains a bottleneck which needs to be solved by batteries, hydrogen and electric vehicles. With the large-scale promotion of electric vehicles, along with the combination of electric vehicles and renewable energy, electric vehicles will become the "real" new energy vehicles utilizing the entire chain of clean energy. In so doing, it could both solve the pollution and carbon emission problems of the vehicle itself, but could also be conducive to the carbon emission reduction of the entire energy system, thus bringing about a new energy revolution for the entire energy system.

Thirdly, intelligent development of system and the AI revolution.

Electric vehicles have three attributes : travel tools, energy devices and intelligent terminals. Intelligent and connected vehicles (ICVs) will restructure the industrial chain and value chain of vehicles. Software defines vehicles while data determine value. The traditional vehicle industry will be transformed into a high-tech industry leading the AI revolution. In the meanwhile, let's take a look at the Internet connection and the feature of sharing regarding vehicles, among "four new attributes", from the perspectives of both the intelligent travel revolution and the new energy revolution: For one thing, the connotation of the Internet attaches equal importance to the Internet of vehicle information and the Internet of mobile energy. For another, the connotation of sharing lays equal emphasis on sharing travel and energy storage information. And both stationery and running electric vehicles can be connected to the mobile energy Internet, finally realizing a full interaction (V2G, Vehicle to Grid). As long as the energy storage scale of distributed vehicles is large enough, it will become the core hub of intelligent transportation energy, namely, the mobile energy Internet. Intelligent charging and vehicle to grid will meet the demand of absorbing renewable energy fluctuations. By 2035, China's inventory of new energy vehicles will reach about 100 million. At that time, the new energy vehicle-mounted battery power will reach approximately 5 billion kW·h (kilowatt-hours) with 2.5 billion – 5 billion kW·h as the charging and discharging power. By 2035, the maximum installed capacity of wind power and photovoltaic power generation will not surpass 4 billion kW. The combination of vehicle-mounted energy storage battery and hydrogen energy could completely meet the demand of load balancing.

All in all, with the accumulation of experience over the past two decades, since 2001, China's electric vehicle industry has shifted to another path and led in the sector of new energy vehicles worldwide. At the same time, China could build its advantage in terms of renewable resources with AI leading the world. It can be predicted that the period between 2020 and 2035 will be a new era when the revolution of new energy electric vehicles, the revolution of renewable energy and the revolution of artificial intelligence will leapfrog and develop in a coordinated manner and create a Chinese miracle featuring the strategic product and industry of new energy intelligent electric vehicles. Focusing on one strategic product and sector, such three technological revolutions and three advantages will release huge power, which could help realize the dream of a strong vehicle nation and play a leading role in all directions. With the help of such advantages, China will create a large industrial cluster with the scale of the main industry exceeding 10 trillion yuan and the scale of related industries reaching tens of trillions of yuan. The development of new energy vehicles at a large scale will result in a new energy revolution, which will bring earthshaking changes to the traditional vehicle, energy and chemical industry, thus truly embracing a great change unseen in a century since the replacement of carriages by vehicles.

The technology revolution of new energy vehicle is advancing the rapid development of related interdiscipline subjects. From the perspective of technical background, the core technology of energy saving and new energy vehicles—the new energy power system technology, remains the frontier technology at the current stage. In 2019, China Association for Science and Technology released 20 key scientific and engineering problems, 2 of them (electrochemistry of high energy and density power battery materials, and hydrogen fuel battery power system) belonging to the scope of new energy system technology; The report of *Engineering Fronts* 2019 published by Chinese Academy of Engineering mentioned the power battery 4 times, fuel battery 2 times, hydrogen energy and renewable energy 4 times as well as electricity-driven/hybrid electric-driven system 2 times. Over the past two decades, China has accumulated plenty of new knowledge, new experience, and so many methods during the research and development regarding new energy vehicles. The research series of key technologies on energy saving and new energy are based on Chinese practice and the international frontier, aiming to review China's research and development achievements on energy saving and new energy vehicles, meet the needs of technological development concerning China's energy saving and new energy vehicles, reflect the key technology research trend of international energy saving and new energy vehicles, and promote the transformation and application of key technologies as regards China's energy saving and new energy vehicles. The series involve four modules: vehicle control technology, power battery technology, motor driving technology as well as fuel battery technology. All those books included in the series are research achievements with the support of National Natural Science Foundation of China (NSFC), major national science and technology projects or national key research and development programs. The publish of the series plays a significant role in enhancing the knowledge accumulation of key technologies concerning China's new energy vehicle, improving China's independent innovation capability, coping with climate change and promoting the green development of the vehicle industry. Moreover, it could contribute to China's development into a strong vehicle nation. It is hoped that the series could build a platform for academic and technological communication and the authors and readers could jointly make contributions to reaching the top in the international stage concerning the technological and academic level in terms of China's energy saving and new energy vehicles.

January 2021

Minggao Ouyang
Academician of Chinese Academy of Sciences
Professor of Tsinghua University (THU)
Beijing, China

Preface

Lithium-ion battery is the most widely used battery in the market. Its main purposes include power battery and energy storage battery. In practical applications, enough cells are usually connected in series to meet the high voltage demand and connected in parallel to meet the high capacity, but each cell in the battery pack is different, which will affect the performance and life of the whole battery pack. Nowadays, in order to avoid inconsistence in the battery pack, different battery-equalization methods are put forward. There are usually two methods of battery equalization, including keeping the state of charge between cells consistent and making the voltage between cells equal. At the same time, the equalization control strategy also includes active cell equalization and passive cell equalization. Compared with the passive equalizing strategies that dissipate energy, active cell equalization method transfers the energy between the cells, which is more efficient and requires shorter equalization time. Because of the advantages of active cell equalization strategies, it has attracted great interest from academia and industries.

In addition, this book introduces the state-of-the-art active cell equalization control strategy for lithium-ion battery packs from the fundamental theories to practical designs. In particular, for different equalization topologies, different equalization control algorithms are used to realize the equalization of each cell in the battery pack. Of course, the same equalization topology can also adopt different equalization control algorithms to realize the equalization of each cell of the battery pack. In addition, charge equalization will also keep the SOC of all cells of the battery pack consistent. This paper mainly includes the following three parts:

- The first part (Chaps. 1–4) first explains that the key role of cell balance is to prolong the service life of battery pack and improve the performance of battery pack; next the battery-equalization system is summarized: the advantages and disadvantages of active equalization system and passive equalization system are compared, and the advantages of active equalization are emphasized, then some new equalizing algorithms of active equalization are introduced. Thirdly, several advanced active cell equalization topology models are introduced. Here, the active equalization topology includes cell-to-cell, cell-to-pack, module-based and layer-based. At the same time, we perform an economic and performance comparison of the above topologies. Finally, according to graph theory, we design the optimal active cell equalizing topology.

- The second part (Chaps. 5–8) designs a neural network-based observer and introduces the three equalization control algorithms based on part of the above topology. For cell-to-cell equalization topology, we first introduce a quasi-sliding mode observer to estimate the SOC value of each cell in the battery pack, then we introduce quasi-sliding mode control-based equalization strategy to realize cell-to-cell equalization, and use hierarchical control to achieve the balance of module-based cell-to-cell topology model. In hierarchical control algorithm, the top layer is the module-level equalizing control, and the bottom layer computes the controlled cell-level equalizing currents for different battery modules in parallel; besides, for cell-to-pack-to-cell equalization control, we divide the battery pack into several modules and introduce an improved module-based CPC equalization, based on the developed model, a two-layer MPC strategy is proposed, where the top-layer MPC controls the ML equalizers and the bottom-layer MPC designs the controlled CMC equalizing currents in each module.
- The third part (Chaps. 9 and 10) designs the two equalization control algorithms based on multi-module charger. For charging strategy of battery pack, the aim is to design a charger so that the SOCs of battery pack converge to the same value. That is, the batteries' SOCs converge to the same desired value in the charging mode. In this part, firstly, we utilize IDA-PBC as bottom-layer controller to make the actual charging currents track their desired values designed by the top layer; secondly, we design a quadratic programming-based simultaneous charging strategy for battery packs, which realize the simultaneous equalization of different battery packs.

Hangzhou, China
November 2022

Jian Chen
Quan Ouyang
Zhisheng Wang

Acknowledgments We would like to sincerely thank Li Gan, Lalitesh Kumar, and Harsh Mohan Sharma for reviewing and editing this book. This work was supported by the Key Research and Development Program of Zhejiang Province, China, under grant 2021C01098, the European Union-funded Marie Skłodowska-Curie Actions Postdoctoral Fellowships, under grant 101067291, and the National Natural Science Foundation of China, under grant 61903189.

Acronyms

AC	Alternating current
AHCM	Ampere hour counting method
AKF	Adaptive Kalman filter
ANNBM	Artificial neural networks-based methods
BMS	Battery management system
CATL	Contemporary Amperex Technology Co., Limited
CC-CV	Constant current-constant voltage
CICM	Continuous input current mode
Cle	Cell-level equalizers
CMC	Cell-to-module-to-cell
CPC	Cell-to-pack-to-cell
CVCS	Constant voltage current sources
DC	Direct current
DICM	Discontinuous input current mode
ECM	Equivalent circuit model
EKF	Extended Kalman filter
EMI	Electromagnetic Interference
EVs	Electric vehicles
FLC	Fuzzy logic control
FLC-BEC	Fuzzy-logic-control battery-equalization controller
FUDS	Federal urban driving schedule
GPIC	General-purpose inverter controller
ICEs	Individual cell equalizers
IDA	Interconnection and damping assignment
MAS	Multi-agent system
MBM	Model-based methods
ML	Module layer
Mle	Module-level equalizers
MOSFET	Metal-oxide-semiconductor field-effect transistor
MPC	Model predictive control
Ni-Cd	Nickel-cadmium

Ni-MH	Nickel-metal hydride
OCV	Open-circuit voltage
OCVM	Open-circuit voltage methods
PBC	Passivity-based controller
PCB	Printed circuit board
PWM	Pulse width modulation
RBFNN	Radial basis function neural networks
RMS	Root mean square
SOC	State of charge
SOH	State of health

Contents

Chapter 1
Introduction

1.1 Applications of Lithium-Ion Batteries

1.1.1 The Crucial Role of Batteries

Extensive research and development on green energy have been completed over the last three decades, yielding a plethora of potential new technologies such as EVs, solar, wind power systems, and telecommunication. These technologies have the potential to reduce reliance on fossil fuels while also alleviating concerns about energy, environmental, and economic issues. The rechargeable lithium-ion battery, as one of the most widely used energy storage devices, plays an important role in the applications depicted in Fig. 1.1 [1].

Because of the wide use of clean energy, global lithium-ion battery market demand is increasing, and it is expected that the demand will gradually increase in the future. The production of lithium-ion battery is shown in Fig. 1.2.

Lithium-ion batteries are generally divided into energy storage batteries and power batteries according to their uses. The energy storage battery is used in photovoltaic or uninterruptible power system, its internal resistance is relatively large, and the charge and discharge speed is relatively slow, generally 0.5–1 C. The power battery is generally used in electric vehicles, its internal resistance is small, and the charge and discharge speed is fast, generally can reach 3–5 C. Usually, the price of power battery is about 1.5 times more expensive than the energy storage battery. Due to the rapid development of electric vehicles, the installed capacity of power batteries is also increasing day by day. In China, the output of power batteries on electric vehicles reached 140 GWh in 2021. The Fig. 1.3 shows the output of power batteries on electric vehicles in recent years. It is estimated that by 2030, the annual output of electric vehicles will reach 20 million. Based on this, we can predict that the output of power batteries will also increase with the increase of the output of electric vehicles.

Of course, lithium-ion batteries can be used as energy storage devices in addition to power batteries. The lithium-ion batteries can store unused or excess electrical energy with a specific lithium-ion battery pack specification, and then extract it for

J. Chen et al., *Equalization Control for Lithium-Ion Batteries*,
https://doi.org/10.1007/978-981-99-0220-0_1

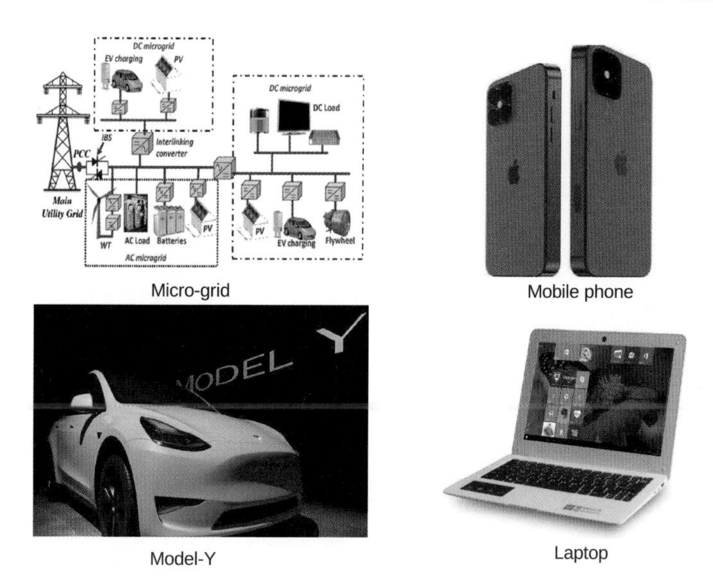

Fig. 1.1 Applications of lithium-ion batteries

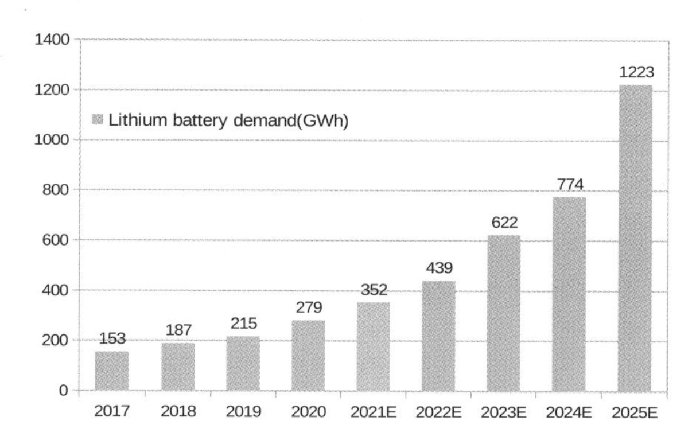

Fig. 1.2 Global lithium-ion battery actual and projected production from 2017 to 2025 [2]

use during peak times, or transport it to the places where reusable energy is scarce. The energy storage system includes the energy storage for households, communications, power grid frequency modulations, and other applications.

It can be seen from the Fig. 1.4 that the existing lithium-ion batteries are mostly used in the field of communications, followed by electric vehicles, and the proportion of household energy storage is very small. However, with the increasing normalization of the use of electric vehicles, it is reasonable to believe that the proportion of household field will increase in the future.

When compared with other energy storage technologies, lithium batteries have numerous incomparable advantages, including very fast response time, high efficiency, low self-discharge, and scaling feasibility due to a modular structure.

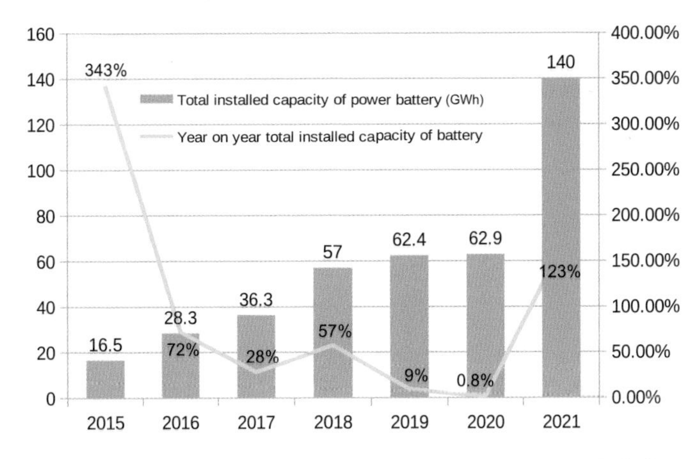

Fig. 1.3 Power battery installed capacity of electric vehicles from 2015 to 2021 [3]

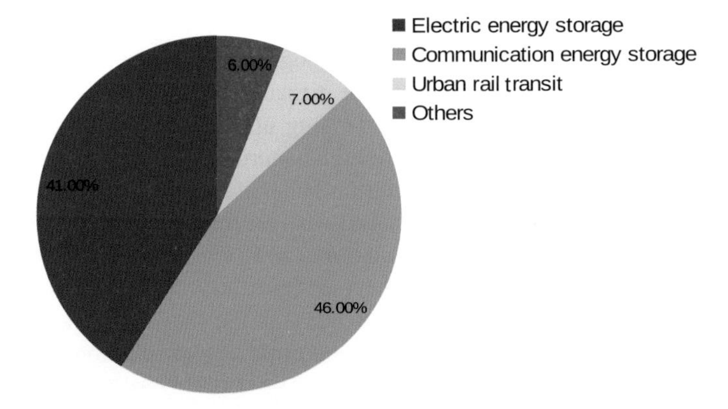

Fig. 1.4 Application fields of China's energy storage battery market in 2020 [4]

In fact, energy storage technology is inextricably linked with renewable energy applications and the development of power grids, which can significantly improve energy efficiency and solve the problems such as power supply in remote areas. Therefore, in order to make renewable energy as the primary source of power, we must improve energy storage technology, which also aids in development of renewable energy technology.

It can be seen from Fig. 1.5 that with the proposal and implementation of new energy technologies, the shipment of energy storage batteries in China will increase to a greater extent in the future. Obviously, with the development of materials and preparation processes, energy storage batteries, cost is getting lower and lower. Meanwhile, the cascade utilization of power batteries' will greatly reduce the cost of the energy storage industry. That is, the rapid development of new energy vehicles will also promote the expansion of the energy storage industry on a large scale.

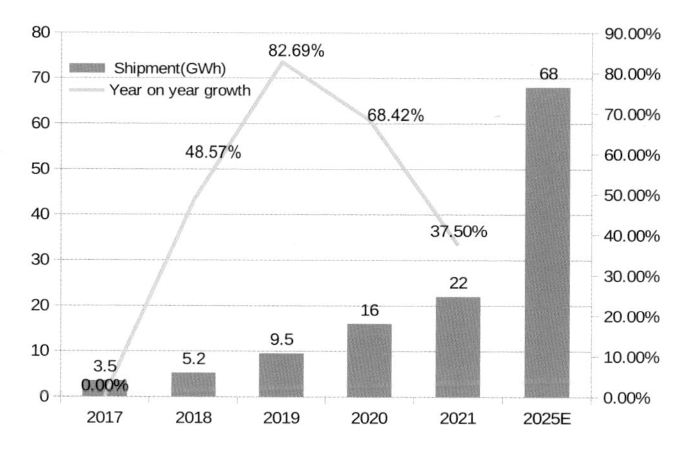

Fig. 1.5 2017–2025 China's lithium battery shipment and its forecast [5]

For lithium-ion batteries, the existing technologies are based on liquid electrolyte. Liquid electrolyte has the disadvantages of difficult storage and low energy density. Therefore, solid-state batteries are stepping up research and development to overcome the shortcomings of existing lithium batteries based on liquid electrolyte. Compared with traditional liquid lithium-ion batteries, solid-state batteries has the following advantages: solid electrolyte has certain mechanical strength and can prevent short circuit to a certain extent. In addition, solid electrolyte has no electrolyte leakage problem, no side reaction at high temperature and no combustion due to gas leakage and it has higher energy density. Because of the above advantages, the current battery academic circles are optimistic about the mass production of solid-state batteries in the near future.

1.1.2 Comparisons of Different Batteries

Currently, the widely used batteries mainly include lead-acid, Ni-Cd, Ni-MH, and lithium-ion batteries [6]. Lead-acid batteries are widely used in automobiles and electric motorcycles because of their mature production technology, high safety, and low price. However, they have a low gravimetric specific power and energy density. The Ni-Cd batteries are ideal DC power suppliers since they have good charge-discharge rate performance. However, they have defects of memory effect and low nominal voltage. Moreover, due to the environmental pollution caused by the heavy metal cadmium, the Ni-Cd batteries are being phased out. Ni-MH batteries have the advantages of a large charge/discharge rate and little memory effect. However, they have a low nominal voltage and are not suitable to be used in parallel. Compared with the above three batteries, rechargeable lithium-ion batteries have the outstanding advantages of excellent design flexibility, low self-discharge rate, high energy density,

Table 1.1 Comparisons of common batteries [7]

	Lead-acid	Ni-Cd	Ni-MH	Lithium-ion
Nominal voltage (V)	2	1.2	1.2	3.2
Gravimetric specific power (Wh/kg)	30–50	40–50	50–70	120–140
Energy density (Wh/L)	60–100	80–100	100–140	240–280
Cycle life (times)	400–600	800–2000	800–2000	1200
Cost (dollars/kWh)	120–150	300–350	150–200	150–180

long lifetime, fast charge capability, and eco-friendly. The comparison results of the above-mentioned batteries are illustrated in Table 1.1.

1.2 Battery Inconsistency Phenomenon

Because of its electrochemical constraints, a single lithium-ion battery cell can only generate voltages ranging from 2.5 V to 4.2 V, which is insufficient to fulfill the high voltage requirement. As a result, in real world applications, a large number of battery cells are generally connected in series to form a high voltage battery pack [1]. Moreover, the battery pack suffers from cell imbalance due to minor differences in the characteristics of each cell. As we all know, the least effective cell limits the battery pack's performance, causing inadequate energy utilization of the whole battery pack. Furthermore, the battery pack's incoherence will cause the batteries to degrade. When batteries are operated in series, they degrade, reducing capacity and even causing a thermal runaway accident during the charging cycle of battery packs.

The Fig. 1.6 shows the consequences of inconsistent batteries. It can be seen that the imbalance of each cell in the battery pack will lead to the consequences of shortening the service life and cause safety accidents of the whole battery pack.

It can be seen from the Fig. 1.6 that thermal runaway will lead to safety accidents such as fire or explosion, which highlights the importance of battery equalization. And the Fig. 1.7 shows the specific fire and explosion caused by thermal runaway of lithium-ion batteries.

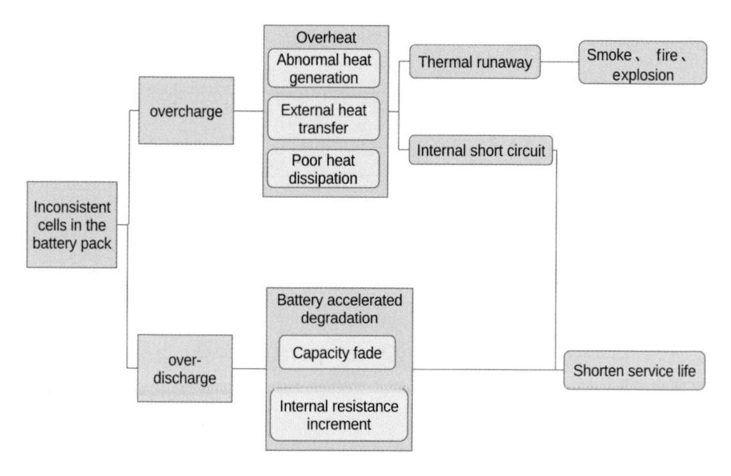

Fig. 1.6 Consequences of inconsistent cells [8]

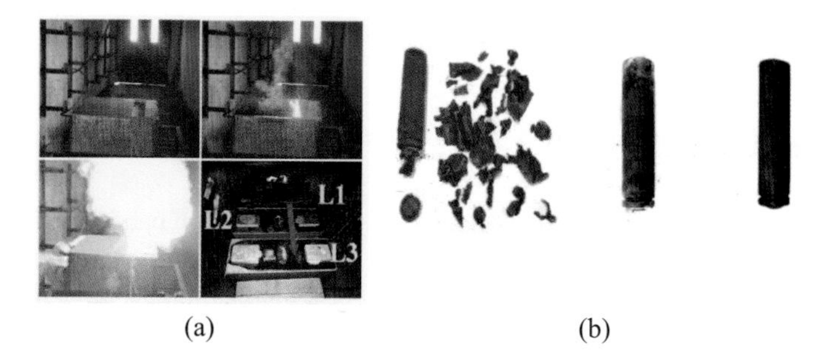

Fig. 1.7 Fire and explosion caused by thermal runaway of lithium-ion batteries [9, 10]

1.3 Crucial Role of Cell Equalization

Because of the serial connection, all cells in the battery pack share the same external charging/discharging current. Even though overcharging and overdischarging can reduce battery pack performance and shorten battery pack lifetime, the charging/discharging process must be stopped when one of the cells is fully charged or discharged. As a result, the cell with the lowest/highest residual available energy limits the overall battery pack's usable/rechargeable capacity. Therefore, if imbalanced cells are used, the battery pack's effective available capacity will decrease. This demonstrates the significance and requirement of cell equalization in the battery pack. At the moment, the main indicators used to assess battery consistency are battery voltage and battery SOC.

1.3.1 Voltage-Based Equalization

Voltage-based equalization aims at maintaining voltages among all cells at the same level. The larger the voltage difference of adjacent cells is, the faster equalizing speed is required. When the difference of adjacent cells reaches a certain criterion,the equalization process stops [11].

The voltage characteristic is the visual expression of the internal resistance of the battery. By comparing the port voltage value between each single cell in the battery pack, if the difference between the two cells meets a certain threshold condition, it is determined that the current state of the battery pack is consistent. Otherwise, the equalization control is triggered to transfer energy to two or more cells whose voltage difference is greater than the threshold. However, due to the inconsistency of the internal resistance of each single cell in the battery pack based on voltage equalization, the inconsistency will gradually intensify after multiple cycles. With the advancement of the technologies, this equalization method has been gradually replaced by State-of-charge based equalization method because of the above said shortcomings.

1.3.2 SOC-Based Equalization

The lithium-ion battery SOC is one of the most important indicators in the lithium-ion BMS, defined as the ratio of the current usable capacity of the battery to its fully charged state capacity [1].

Since there is an identified relationship between the open circuit voltage and the SOC of the battery cell, numerous cell balancing algorithms are designed for cells' OCV or voltage equalizing issues in the literature. However, a large difference in cell's SOC can be projected on a small bias in the cell's OCV since the cell's OCV is almost a constant in a typical SOC operation range (20%–90%). The SOC-based cell equalization will be more beneficial to the battery pack's performance, and thus, preferred over the voltage equalizing. Therefore, the aim of cell equalization is to maintain the cells' SOC differences at the same level. The maximum allowed cell equalizing current is designed to vary with the change of the external current of the battery pack rather than a constant to avoid the cells' currents exceeding their limitations. In general, compared with voltage-based equalization, SOC-based equalization has two advantages: (1) This will make the model easier to be applied to equalization systems of different types of batteries; (2) it does not need to consider the nonlinear voltage of the battery, which will make the equalization system algorithm (expressed in state-space) easier to implement. Therefore, SOC-based equalization is the mainstream of the equalization mode (Table 1.2).

Usually, the cells' SOC equalization mainly includes the following two steps: (1) prior to the cell balancing process, the designed observer estimates the cells' SOC

Table 1.2 Comparisons of two cell equalization methods

	Voltage-based equalization	State-of-charge-based equalization
Advantages	Easy operation	Higher equalization performance
Disadvantages	Poor equalization performance	Poor operability and complex structure
Applications	Battery pack with fewer cells	Battery pack with more cells

and the cell equalizing controller remain deactivated; (2) when the cells' SOC estimation errors become small enough, the cell balancing controllers are turned on [12].

Inaccurate battery SOC estimations can easily cause overdischarging or overcharging of batteries, resulting in irreversible damage to the battery, such as fire, explosion and etc. Hence, the SOC estimation is very important in the process of SOC-based equalization. In order to estimate the SOC of lithium-ion batteries under different operating conditions accurately, a model describing the characteristics of the battery is mandatory. Commonly used lithium-ion battery models include electrochemical models [13–15], neural network model [16, 17] and equivalent circuit model [18, 19].

The electrochemical reaction process inside a cell can be described by the partial differential equations, and is often used to describe the dynamics and electrode/electrolyte material selection of lithium-ion batteries. The parameters in the battery model are dependent on the factors such as the structure, size, and material of the battery. Therefore, the electrochemical model of the battery becomes very complex, making it difficult to use the model for SOC estimation.

The complex dynamical model of lithium-ion battery can be obtained by using the self-learning technique of the neural networks, which also exhibits strong nonlinearity, and helps in simulating the complex dynamics. The neural network can determine the dynamical characteristics without knowing the internal reaction mechanism of lithium-ion battery. However, the neural networks require a large amount of training data, the amount of computation is large, and the training effect cannot be guaranteed.

The equivalent circuit model uses circuit components to simulate the dynamic characteristics of lithium-ion batteries, with low computational complexity and moderate model accuracy. What's more, the equivalent circuit model can be easily extended to other types of batteries. Therefore, compared to the other two models, the equivalent circuit model is best suited for lithium-ion battery SOC estimation models.

For lithium-ion battery SOC estimation, the online SOC estimation algorithm can be divided into four categories according to the principle: AHCM, OCVM, MBM and ANNBM. Among them, the AHCM needs to know the initial value of the SOC, and the lithium-ion battery will cause the initial value to be unknown if it is not used continuously, [20] proposes a method to improve the accuracy of online estimation of the charging efficiency of lithium batteries. The open circuit voltage and SOC in the OCVM are affected by the temperature and service time of the lithium-ion battery [21–24].The traditional OCVM are difficult to apply in most online scenarios. The MBM rely on precise battery models for accurate SOC estimation [25, 26]. However, the internal parameters change during battery charge and discharge, and it is difficult to establish an accurate model to describe the external characteristics of all batteries sufficiently, especially the computational complexity of the battery model should be limited to a reasonable range for online applications [27]. Presence is insensitive to initial SOCs, robust to measuring noise, and MBM are very popular in various online SOC estimations. The ANNBM can directly establish the relationship between SOC and other variables such as current, voltage, and temperature [28–30] without understanding the mechanism model inside the lithium-ion battery, if the appropriate sample is selected and the parameters are optimized during training, the artificial neural network system can provide an accurate SOC estimation.

On the whole, although the AHCM is simple to implement, measurement errors could accumulate in real-world applications due to its open-loop design. EKF and AKF need to linearize the nonlinear battery system (the nonlinear relationship between battery OCV and SOC), the accuracy of SOC estimation will be reduced. However, to the best of our knowledge, the SOC estimation performance of above methods cannot be guaranteed as their stability has not been proved and no convergence analysis has been given related to the actual battery system. Since the process and the measurement noise are assumed to be Gaussian white noise in these algorithms, their performance will be reduced in situations when the noise covariance cannot be properly obtained.

Based on the nonlinear observer, an RBFNN is utilized to estimate the battery uncertainties online. It can be proved that the bound of the SOC estimation error is arbitrarily small. So it can be obtained that the SOC of the battery can be estimated by the RBFNN-based nonlinear observer accurately [31].

With good ability to overcome the uncertainties and disturbances of the system, quasi-sliding mode-based strategies can be good solutions for the cell's SOC equalization in the battery pack. Besides, it can be concluded that the SOC estimation-based quasi-sliding mode control algorithm has the more excellent performance for the cell equalization of the serially connected battery pack [12].

Above all, the RBFNN and quasi-sliding mode-based state observer has higher accuracy and faster convergence speed than others (Table 1.3).

Table 1.3 Comparisons of common SOC estimation algorithms

	Amphere-hour	EKF	RBFNN	Sliding-mode
Advantage	Easy to implement	Easy to implement	Higher accuracy	Higher accuracy
Weakness	Time consuming	Lower accuracy	Complex structure	Complex structure

References

1. N. Ghaeminezhad, Q. Ouyang, X. Hu, G. Xu, Z. Wang, Active cell equalization topologies analysis for battery packs: a systematic review. IEEE Trans. Power Electron. **36**(8), 9119–9135 (2021)
2. F.I.R. Institute, Report of market demand forecast and investment strategy planning on China li-ion power battery industry (2021–2026), (Foresight Industry Research Institute, Technical Report, China, 2021)
3. C. Zhao, Review of domestic power battery market, (GGII, Technical Report 2021)
4. F.I.R. Institute, China's energy storage lithium battery application fields in 2020, (Foresight Industry Research Institute, Technical Report, China, 2021)
5. C. Zhao, Research and analysis report on China's energy storage lithium battery market in 2021, (GGII, Technical Report 2021)
6. Q. Ouyang, Research on key technologies of lithium-ion battery management system for electric vehicles. Ph.D. Dissertation, Zhejiang University, 2018
7. S. Zhang, Research on equalization algorithm based on LiFePO4 cell state of charge. Ph.D. dissertation, Shanghai Jiaotong University, 2015
8. M.H. Lipu, M. Hannan, T.F. Karim, A. Hussain, M.H.M. Saad, A. Ayob, M.S. Miah, T.I. Mahlia, Intelligent algorithms and control strategies for battery management system in electric vehicles: Progress, challenges and future outlook. J. Clean. Prod. **292**, 1–27 (2021)
9. T. Zhou, C. Wu, B. Chen, H. Zhu, Y. Liu, Fire suppression and cooling effect of perfluorohexanone on thermal runaway of lithium-ion batteries with large capacity, in *2021 IEEE 5th Conference on Energy Internet and Energy System Integration (EI2)*, Taiyuan, China, 2021, pp. 3783–3788
10. F. Jiang, W. Wang, Y. Chen, D. Liang, S. Mo, Failure and microstructure characteristics of lithium batteries under different overcharging voltage conditions, in *2019 9th International Conference on Fire Science and Fire Protection Engineering (ICFSFPE)*, Chendu, China, 2019, pp. 1–5
11. J. Zheng, J. Chen, Q. Ouyang, Variable universe fuzzy control for battery equalization. J. Syst. Sci. Complex. **31**(1), 325–342 (2018)
12. Q. Ouyang, J. Chen, J. Zheng, Y. Hong, Soc estimation-based quasi-sliding mode control for cell balancing in lithium-ion battery packs. IEEE Trans. Ind. Electron. **65**(4), 3427–3436 (2018)
13. Y. Wang, H. Fang, Z. Sahinoglu, T. Wada, S. Hara, Adaptive estimation of the state of charge for lithium-ion batteries: Nonlinear geometric observer approach. IEEE Trans. Control. Syst. Technol. **23**(3), 948–962 (2014)
14. M. Corno, N. Bhatt, S.M. Savaresi, M. Verhaegen, Electrochemical model-based state of charge estimation for li-ion cells. IEEE Trans. Control. Syst. Technol. **23**(1), 117–127 (2014)
15. S. Dey, B. Ayalew, P. Pisu, Nonlinear robust observers for state-of-charge estimation of lithium-ion cells based on a reduced electrochemical model. IEEE Trans. Control. Syst. Technol. **23**(5), 1935–1942 (2015)
16. J. Wu, Y. Wang, X. Zhang, Z. Chen, A novel state of health estimation method of li-ion battery using group method of data handling. J. Power Sources. **327**, 457–464 (2016)
17. L. Kang, X. Zhao, J. Ma, A new neural network model for the state-of-charge estimation in the battery degradation process. Appl. Energy. **121**, 20–27 (2014)
18. R. Relan, Y. Firouz, J.M. Timmermans, J. Schoukens, Data-driven nonlinear identification of li-ion battery based on a frequency domain nonparametric analysis. IEEE Trans. Control. Syst. Technol. **25**(5), 1825–1832 (2016)
19. M. Chen, G. Rincon-Mora, Accurate electrical battery model capable of predicting runtime and I-V performance. IEEE Trans. Energy Convers. **21**(2), 504–511 (2006)
20. K.S. Ng, C.S. Moo, Y.P. Chen, Y.C. Hsieh, Enhanced coulomb counting method for estimating state-of-charge and state-of-health of lithium-ion batteries. Appl. Energy. **86**(9), 1506–1511 (2009)
21. M.-W. Cheng, Y.S. Lee, M. Liu, C.C. Sun, State-of-charge estimation with aging effect and correction for lithium-ion battery. IET Electr. Syst. Transp. **5**(2), 70–76 (2015)

22. S. Tong, M.P. Klein, J.W. Park, On-line optimization of battery open circuit voltage for improved state-of-charge and state-of-health estimation. J. Power Sources. **293**, 416–428 (2015)
23. Y. Xing, W. He, M. Pecht, K.L. Tsui, State of charge estimation of lithium-ion batteries using the open-circuit voltage at various ambient temperatures. Appl. Energy **113**, 106–115 (2014)
24. B. Pattipati, B. Balasingam, G. Avvari, K. Pattipati, Y. Bar-Shalom, Open circuit voltage characterization of lithium-ion batteries. J. Power Sources. **269**, 317–333 (2014)
25. C. Zhang, L. Wang, X. Li, W. Chen, G.G. Yin, J. Jiang, Robust and adaptive estimation of state of charge for lithium-ion batteries. IEEE Trans. Ind. Electron. **62**(8), 4948–4957 (2015)
26. X.W. Yan, Y.W. Guo, Y. Cui, Y.W. Wang, H.R. Deng, Electric vehicle battery SOC estimation based on GNL model adaptive kalman filter. J. Phys. Conf. Ser. **1087**(5), 052027 (2018)
27. M. Cacciato, G. Nobile, G. Scarcella, G. Scelba, Real-time model-based estimation of soc and soh for energy storage systems. IEEE Trans. Power Electron. **32**(1), 794–803 (2016)
28. J.C. Alvarez Anton, P.J. Garcia Nieto, C. Blanco Viejo, J.A. Vilán Vilán, Support vector machines used to estimate the battery state of charge. IEEE Trans. Power Electron. **28**(12), 5919–5926 (2013)
29. L. Zhao, M. Lin, Y. Chen, Least-squares based coulomb counting method and its application for state-of-charge (SOC) estimation in electric vehicles. Int. J. Energy Res. **40**(10), 1389–1399 (2016)
30. X. Hu, F. Sun, Y. Zou, Comparison between two model-based algorithms for li-ion battery SOC estimation in electric vehicles. Simul. Model. Pract. Theory. **34**, 1–11 (2013)
31. J. Chen, Q. Ouyang, C. Xu, H. Su, Neural network-based state of charge observer design for lithium-ion batteries. IEEE Trans. Control. Syst. Technol. **26**(1), 313–320 (2017)

Chapter 2
Overview of Cell Equalization Systems

2.1 Classification and Comparisons of Cell Equalization Systems

The inconsistency between individual cells is gradually amplified as cycle times increase, affecting the overall performance of the battery pack. In general, equalization technology is used to incorporate battery pack equalization control in order to extend service life while also ensuring the better performance. At present, the research on battery equalization technology is mainly divided into two parts: One part is the design of equalization circuit topology; the other part is the research of equalization control strategy. For the design of equalization circuit topology, it can be divided into dissipative equalization circuit and non-dissipative equalization circuit from the perspective of energy. Generally, these two equalization methods are also called passive equalization and active equalization. Absolutely, the advantages and disadvantages of the two equalization methods and the applications are also compared in this section.

2.1.1 Passive Cell Equalization Systems

Shunt resistors [1–3] are used in passive cell balancing to dissipate excess energy from cells with high SOC/voltage in order to balance cells with lower SOC/voltage [4]. Even through its advantages in terms of simplicity, low cost, and reliability, passive cell balancing's traditional energy dissipation reduces the available capacity of the battery pack and necessitates the use of an additional heat management system [5, 6]. At present, there are two main methods based on dissipative equalization topology, as shown in Fig. 2.1. One is fixed resistance equalization, the other is controllable resistance equalization.

J. Chen et al., *Equalization Control for Lithium-Ion Batteries*,
https://doi.org/10.1007/978-981-99-0220-0_2

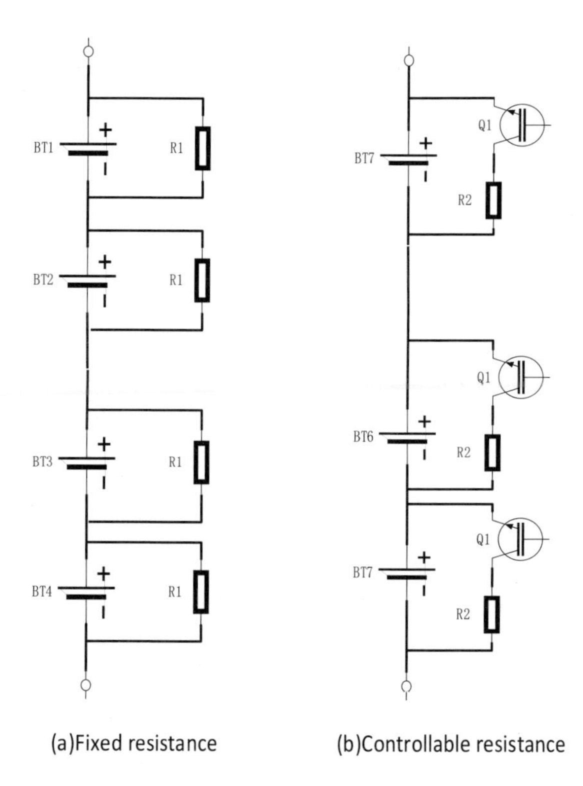

Fig. 2.1 Schematic diagram of two kinds of passive equalization [7]

The Fig. 2.1a shows the fixed resistance equalization mode. The BMS system does not participate in the equalization management of the battery pack, but only monitors the voltage and temperature of the single battery. The single battery continuously discharges the single resistance, and the discharge resistance can be selected according to the specific state of the single battery. Under any working condition, the radiation type has the advantages of low heat consumption and high reliability.

The Fig. 2.1b shows the controllable resistance equalization mode. The BMS system controls the controllable devices to turn on or off by monitoring the port voltage of each single battery to balance the energy. In the charging condition, the opening time of the controllable device is controlled in the form of duty cycle to realize the shunt regulation of the single battery that needs to be balanced, and the shunt equalization of multiple batteries can be realized at the same time.

To sum up, the advantages of resistive equalization control are simple hardware implementation scheme and low cost. At present, the new energy vehicles in the market including Tesla adopt this balanced approach. Of course, the disadvantages of this method are also obvious. It realizes balance by consuming excess power and converts electric energy into heat energy, which not only reduces the usable efficiency of energy, but also poses a severe challenge to the PCB thermal management of BMS.

2.1.2 *Active Cell Equalization Systems*

The movement of charge from high SOC/voltage cells to low SOC/voltage cells can be made possible by incorporating active equalization circuits in active cell equalizing technique [8–11]. In active approach of cell equalization, the energy transfer between the cells can be done more efficiently with shorter equalization time as compared to passive equalization approach [12–14]. Furthermore, the active cell balancing technique has several advantages that have piqued the interest of researchers [15–20].

The active cell equalization circuit topology can be divided into switching capacitor equalization, inductance equalization, LC (inductance and capacitance) equalization, and transformer equalization circuits according to the different energy storage components [21, 22]. The key to the design of active equalization circuit lies in the balance of structural complexity, system cost and equalization efficiency. Its ultimate goal is to design the optimal circuit structure with higher equalization efficiency with less system cost.

Capacitance equalization [23–26] uses capacitor for energy transfer. The switching capacitors in the circuit are generally turned on and off, and the energy is transferred from high SOC/voltage to low SOC/voltage batteries through capacitors to achieve equalization. Although its structure is simple, it has the disadvantages of low efficiency and high energy consumption, also when the voltage difference between the battery cells is small, the equalization speed is very slow. This method has long equalization time compared to other active methods [27].

In some studies [28, 29], energy transfer had been accomplished by utilizing inductance. The control of equalizing current is not difficult in the case of inductance equalization, but equalizer performance can be determined by the circuit structure and equalization approach of the equalizer. It has faster equalization time and lower cost compared with capacitor equalizing [30, 31].

SOC equalization is realized by an LC oscillation circuit [32–34], that further indemnifies for the shortcoming of the small voltage difference in capacitance equalization. However, it has a lower equalization efficiency, a higher switching frequency, and requires complex equalization circuit control. LC equalization circuit includes Cûk converter, flyback converter, quasi-resonant zero converter and buck-boost converter. Among them, Cûk converter is the typical converter of the LC equalization circuits. The Cûk converter equalizing circuit has higher efficiency with non-dissipative currents and a bidirectional energy transferring capability; flyback converter has simple structure, simple control, high reliability and isolation, but the output power is limited. It is mostly used in circuits below 200W; quasi-resonant converter method drives the switch tube to adjust the voltage of single battery through PWM signal, which not only improves the equalization efficiency, but also effectively reduces the switching loss and EMI electromagnetic radiation. It is suitable for medium power level, but the circuit is complex and the cost is too high; for the buck-boost circuit equalization topology, only one equalization inductor is used, which reduces the structural complexity and improves the equalization speed. However, the premise of

Table 2.1 Comparisons of six different equalization devices

Name	Components	Equalizing speed	Control difficulty	Cost	Size	Efficiency
Resistor [7]	n(SW)+n(R)	Slow	Easy	Low	Small	Low
Switched capacitance [25]	2n(SW)+n(C)	Slow	Difficult	Low	Medium	Low
Switched inductance [29]	n(SW)+n(C)	Slow	Difficult	Low	Medium	Low
Buck-boost converter [39]	(n+2)(L)+(2n+4)(SW)	Fast	Difficult	High	Medium	High
Cûk converters [40]	n(SW+D)+n(L)+(n/2)(C)	Fast	Difficult	High	Large	High
Transformer [36]	2n(SW)+1(T)	Fast	Difficult	High	Large	Low

n is the number of the batteries in the battery, switch (SW), resistance (R), capacitance (C), inductance (L), diode (D),transformer(T)

its two-way equalization control is the increase of a large number of switches, which brings the deficiency of the increase of system cost and control complexity.

Transformer equalization [35, 36] mostly adopts the multi-output isolation transformer structure, the primary side is connected with the positive pole of the battery, and the secondary side of the multi-winding is respectively connected with each single battery. It has fast equalization speed but because of nonuniform turn-ratio of secondary winding leakage inductance, the voltages of the secondary windings are not equal [37, 38]. The battery with large single battery voltage has large charging current, on the contrary, the battery with small voltage has small charging current and small energy, so as to realize equalization. This topology has the advantages of low loss and fast efficiency, but its structure is complex.

To summarize, previous equalization network architecture circuits have issues including sluggish equalization, poor controllability of equalization energy, and inadequate equalization implications. Although many studies on equalization control algorithms have been conducted and adequate equalization effect has been accomplished, it frequently keeps failing to satisfy the expectation for equalization speed in industrial applications (Table 2.1).

2.1.3 Comparisons of Cell Equalization Systems

The specific advantages and disadvantages of the two equalization circuit topologies are compared. The results show that the passive equalization system has the advantages of simple structure, low cost, stability and reliability; it's disadvantages are slow equalization speed and high energy consumption. And the active equalization system's advantages are their equalization speed is fast, and there is basically no energy dissipated for no reason; it's disadvantages are complex structure, high cost and high failure rate. To sum up, active equalization system has higher energy efficiency and shorter equalization time than passive equalization system.

Although the active equalization circuit has many outstanding advantages compared with the passive equalization circuit, despite some obvious limitations, such as heat dissipation, a time-consuming process, and energy waste passive equalization is the most popular technique of equalization today. Active equalization technology is still unmature and can not be better applied to the commercial field. However, with the increasing exposure of active equalization in BMS applications, it will occupy a certain market in commercial applications in the future.

2.2 Commercial Equalizers

Many existing works incorporate active equalization techniques in hardware circuits that use inductors and switches to transfer energy. Such active equalization technique implementations are shown in, Fig. 2.2a–c for Cûk, buck-boost, and quasi-resonant converters, respectively. These equalizers are used to transfer energy from a higher to a lower SOC level between two adjacent cells. It can be seen that the structural symmetry of the converter regulates the bidirectional energy flow between the connected cells. Some more active equalizers are shown in Fig 2.2d–g, depicting bidirectional type multi-secondary windings transformer, bidirectional type multiple transformers, bidirectional type flyback converter, and bidirectional type switched transformer, respectively. These transformers are utilized for balancing the charge between the adjoining cells as well as between the cells and the pack. In some cases, the charge transfer does not restricted to the adjoining or non-adjoining cells but also from cell to cell directly. An example is shown in Fig. 2.2h which is the hardware circuit application of single switched capacitor circuit.

2.2.1 *Bidirectional Buck-Boost Converters*

The isolated and customized bidirectional type buck-boost converters can be used as the ICEs, but the proposed results can easily be extended to other types of ICEs.

Operational principle: An illustration of isolated and modified bidirectional type equalizer consisting of buck-boost converter is shown in Fig. 2.3a. The subscript l represents the ICE composed of a transformer T_l, two resistors R_{l1} and R_{l2}, and two MOSFETs Q_{l1} and Q_{l2}. The symmetrical structure of this ICE allows the transfer of energy between the i-th and the j-th connected cells in bidirectional mode. The balancing procedure could be divided into different steps the i-th (j-th) cell to the j-th (i-th) cell for the transfer of energy in a particular switching period:

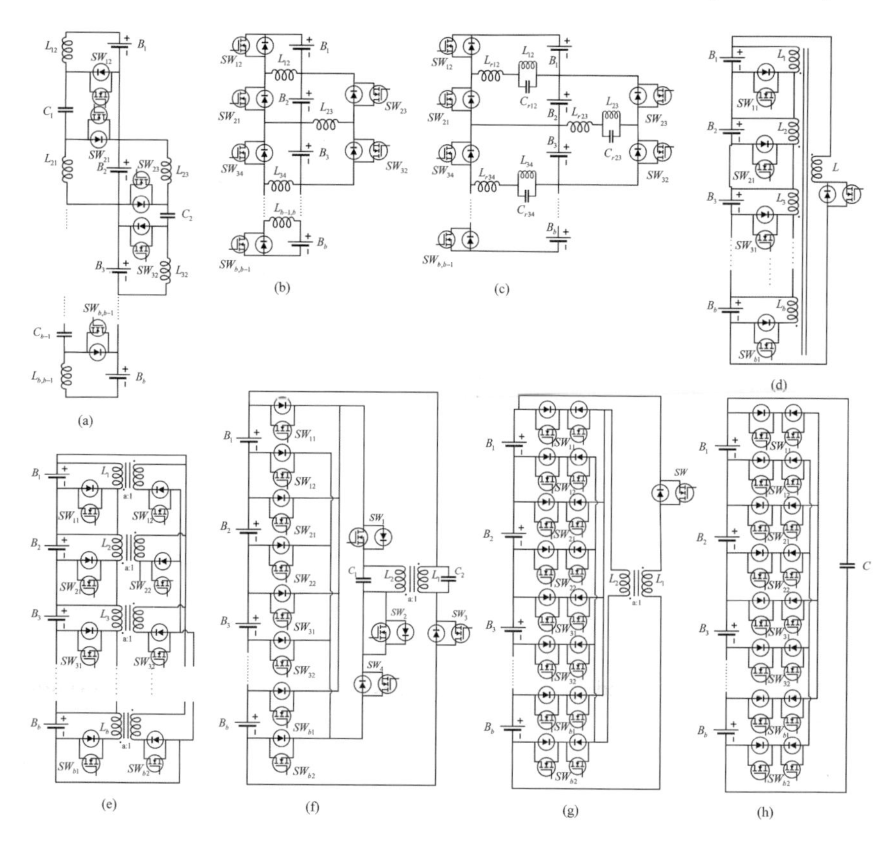

Fig. 2.2 Hardware circuits of the active cell equalizers **a** CûK converter, **b** buck-boost converter, **c** quasi-resonant converter, **d** bidirectional multi-secondary windings transformer, **e** bidirectional multiple transformers, **f** bidirectional flyback converter, **g** bidirectional switched transformer, **h** single switched capacitor [41]

- Step 1: MOSFET Q_{l1} (Q_{l2}) is turned on, and MOSFET Q_{l2} (Q_{l1}) is turned off. Transformer T_l is charged by the i-th (j-th) cell.
- Step 2: MOSFET Q_{l1} (Q_{l2}) is turned off, and MOSFET Q_{l2} (Q_{l1}) is turned on. Transformer T_l charges the j-th (i-th) cell.

The described steps above are repeated to transfer the energy from the i-th (j-th) cell to the j-th (i-th) cell. The inductor volt-second balance principle proposed in [42] says that the average current in inductors should be kept constant while operating the converter in steady-state. Therefore, the average current in inductors can be used as the equalizing currents.

Example 2.1 The parameters of the circuit are selected as $R_{l1} = R_{l2} = 25\ \text{m}\Omega$ and $T_l = 10\ \mu\text{H}$ for connecting the i-th and j-th cells of l-th buck-boost converter which is isolated and modified to measure the equalizing currents by utilizing the selected resistors. The voltages of the i-th and j-th cells are set as $V_{B_i} = 3.91$ V and

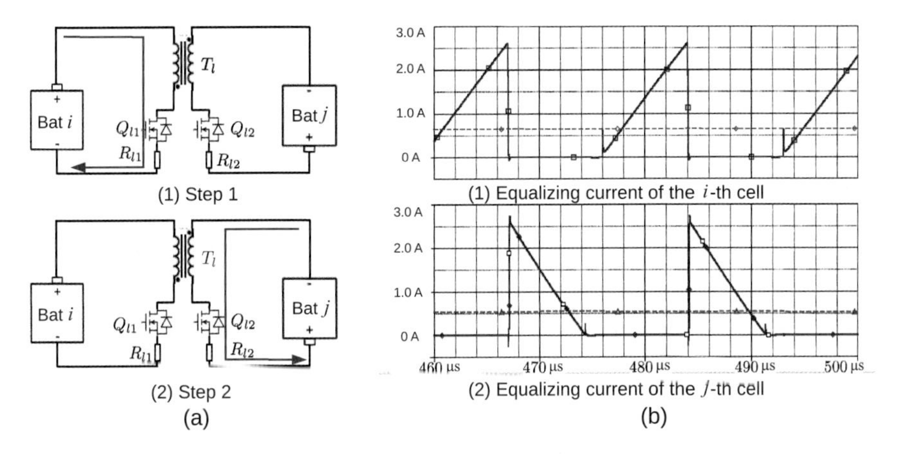

Fig. 2.3 **a** Balancing of two cells by an isolated modified buck-boost converter, **b** Equalizing currents supplied by the *i*-th and to the *j*-th cells

$V_{B_j} = 3.72$ V, respectively. The transient period of the equalizing current curves passing through the bidirectional type isolated buck-boost converter is shown in Fig. 2.3b as an illustration. The simulations of equalization process are performed by Cadence Dspice tool. Their average equalizing currents are about 0.63 A and 0.54 A, respectively. The efficiency of energy transfer is approx. 85% through the l-th converter, which is consistent according to the analysis of isolated and modified buck-boost converter.

2.2.2 Bidirectional Modified Cûk Converters

For cell-to-cell balancing circuits having the fast equalization, excellent implementation, small size, and low cost, one of the most commonly used converters is bidirectional type and modified Cûk converter. Other benefits include modular design and an integrated infrastructure for battery equalization applications.

DICM can also support bidirectional and modified Cûk converters. The MOSFETS are turned ON while keeping the body diodes turned OFF at zero current state to start the DCIM operation, this can decrease the power loss in MOSFETs switch significantly and also improve the average balancing efficiency as compared with those designed in CICM.

As depicted in Fig. 2.4, the dynamical response of i-th converter $(1 \le i \le n-1)$ obtained through DCIM operation can be separated into three different states in one switching period. The i-th and $(i+1)$-th battery cells are connected with the i-th converter. The converter consists of two uncoupled inductors L_{i1} and L_{i2}, two MOSFETs Q_{i1} and Q_{i2} with body diodes d_{i1} and d_{i2}, and a capacitor to transfer the energy C_i. The converter circuit is regulated by the PWM signals, which is

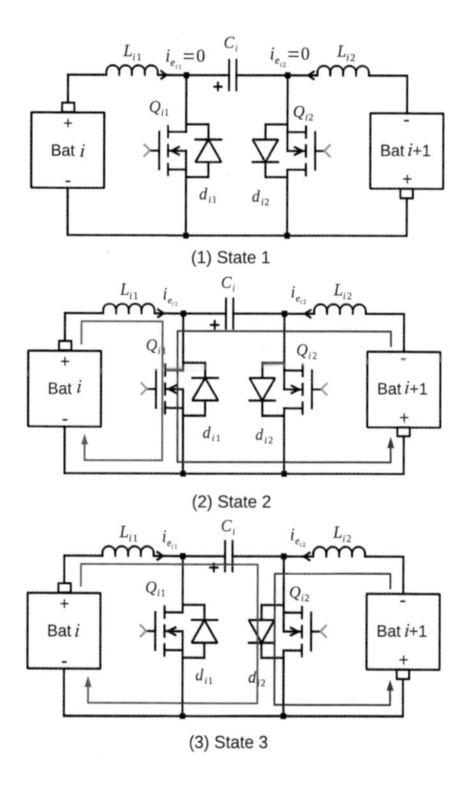

Fig. 2.4 The periodic dynamics of the converter in DICM

responsible to control the switching on and off of the MOSFETs' to accomplish the cell equalization. The duty cycles D_{i1} and D_{i2} of the PWM signals are to be selected in order to control the variables for cell equalization. To minimize the MOSFETs' switching loss, the converters are designed in such a way that it can be operated in the DICM mode [43].

Due to symmetrical structure of the bidirectional converter the transfer of charge can take place from the i-th cell to the $(i+1)$-th cell. The dynamic response of the converter is divided into three states in one time period T_s for the DICM operation, as follows [44]:

- State 1: The current is not flowing through the converter as detailed above, which means $i_{e_{i1}} = i_{e_{i2}} = 0$.
- State 2: MOSFET Q_{i1} is on. Then cell i discharges to L_{i1}, and L_{i1} stores the energy. The capacitor C_i discharges to cell $i+1$ and L_{i2}, and L_{i2} stores energy.
- State 3: MOSFET Q_{i1} is off and d_{i2} is forced to turn on. Then cell i discharges to L_{i1} and capacitor C_i, and both L_{i1} and C_i store energy. Moreover, L_{i2} discharges to cell $i+1$.

When the current approaches 0, the converter comes to the initial state. A battery equalization loop contains the three states described above.

2.3 Overview of Equalization Algorithms

Cell equalization can be obtained in many ways such as from cell-to-cell and cell-to-pack-to-cell [41], based on these two topologies, many scholars have proposed different algorithms to achieve the balance between batteries. In addition, charge equalization is also an important type of active equalization.

2.3.1 Cell-to-Cell Equalization Algorithms

There are numerous effective cell balancing techniques available in the literature that propose cell-to-cell equalization topology. To regulate the equalizing current between two adjacent cells for transferring the energy and achieving cell equalization, an intelligent control based on fuzzy concept is propounded in [4]. For balancing the voltage of cells, a fuzzy logic based controller is proposed in [44] by operating the converters in DICM mode [45]. In [46], it was elaborated in detail that how the fuzzy based controller changes the input by adapting the equalizing current, which was accomplished with suitable rule bases and influence of the input-output signals. However, if the change in inputs is very small in a large universe because the cell imbalance is associated with inputs, then the number of rule base is sufficiently large to adapt the minor change in inputs [47]. In practice, it is very complex and tedious to accomplish this minor change in inputs. In order to tune the equalizing current and to improve the equalization speed for different conditions [48], a variable universe controller design is propounded based on the fuzzy concept. The variable universe is defined as the input universes of discourse, which signifies the change in output variables as the input variable changes. The concept of variable universe in designing the controller with fuzzy framework is more effective than the conventional fuzzy based controller because of the better efficiency and adaptivity of the variable universe controller based on fuzzy concept. Because of the high non-linearity content in the battery model, incorporating the concept of variable universe in fuzzy framework is one of the options for battery equalization. To minimize the switching loss in the MOSFETs, Ouyang has incorporated modified bidirectional Cûk converters operating in DICM mode. For tuning the duty cycles of PWM signals during the equalizing process, $(n-1)$ online voltage estimation based variable universe controllers designed in fuzzy framework are proposed for a n cell serially connected battery pack. The proposed variable universe fuzzy controller can handle the equalization problem effectively without creating an excessive number of rules for the fuzzy set. Besides this, during the balancing process, adjoining cells either with excessively low or excessive voltage cannot sustain large currents. A neuro-fuzzy controller is propounded in [12] with experimental validation for achieving the cell equalization. The neurons have the self-learning capability to tune the controller parameters adaptively. A nonlinear MPC is propounded in [49] for balancing the

SOC of cells' and an optimal solution of equalizing currents is obtained by genetic algorithm optimization.

In the existing literature, to the best of our knowledge, the consistency of cells' SOC for the battery pack is not investigated in equalizing control technique, and the balancing techniques discussed above are only useful for equalizing two adjacent cells. Furthermore, the effect of the external current of the battery packs on the equalizing currents has not been taken into account in the literature. Since the excess of equalizing currents during battery charge/discharge with large external current is not possible. The ICEs are frequently used in cell-to-cell balancing topologies. Their popularity can be attributed to their fast equalization, ease of operation, reasonable size, and cost. Furthermore, because they permit for integrated infrastructure and design that allows, they are appropriate for a broad range of real world applications. The constrained optimization for simultaneous thermal and SOC balancing in [6], the SOC-based droop method in [9], the adaptive backward control algorithm in [12], the modular equalization technique in [8] are some of the available investigations on the ICEs based topologies for cell-to-cell balancing. Sliding-mode-based techniques are known for their fast convergence and impeccable robustness due to their ease of implementation. Additionally, the sliding mode-based designs can overcome the perturbations and uncertainties that are injected/inherently present in the system. In [50], a discrete-time quasi-sliding-mode assisted cell balancing methodology with saturated equalizing current constraints is proposed to have the bidirectional modified Cûk converters cooperate efficiently for the realization of the cells' SOC equalization.

Researchers have shown an interest in distributed consensus control techniques for re-configurable battery systems that provide access to individual cells for monitoring and control in recent years [51]. A cell-to-pack-to-cell topology is propounded by utilizing multi-agent control technique for cells equalization of the battery pack [52]. A distributed control algorithm based on multi-agent concept has been propounded in [53] for SOC equalization of distributed energy storage units in an AC micro-grid. The decentralized balancing approaches propounded in [4, 12, 46, 49, 54] are based on a decentralized structure which controls the equalizer individually for balancing its two adjacent cells. Due to lack of co-ordination between the equalizer control algorithms, the equalizing speed decreases in the cell balancing system. In [54], the authors have investigated an interesting problem to address the speed of equalization procedure in significant amount by incorporating some ICEs. It can be demonstrated that the balancing time decreases as the second smallest eigenvalue of the Laplacian matrix of the cell/ICE based MAS increases, which quantifies the MAS's algebraic interconnection. The question then becomes how to effectively determine the ICE edges added to the original MAS to maximize algebraic connectivity. An optimization problem can be structured in this manner and solved using 0-1 programming.

So that each cell equalizer can effectively work together to achieve the equivalent of each monomer lithium-ion battery [55], where the maximum allowable equivalent current of the battery is designed to be a variable that changes with the external current of the battery pack, so as to avoid the battery current exceeding its allowable range. Further, a module-based cell balancing system was designed, which includes cell balancing and cell-to-module balancing. Considering the two balancing goals of battery pack balancing time and temperature, as well as the battery charge state, current and equivalent current constraints, a multi-target constraint optimization problem for the overall coordinated control of all equalizers is constructed. In order to reduce the amount of calculation to facilitate the solution, a hierarchical equalization control method is proposed, first of all, the top layer controls the balanced current between the battery modules, and the bottom layer controls the equivalent current between the cells in each battery module in parallel. The hierarchical control framework designed significantly reduces the amount of computation.

2.3.2 Cell-to-Pack-to-Cell Equalization Algorithms

The equalization system with cell-to-cell topology uses balancers to transfer energy between the adjacent cells directly. Although it has a simple modular structure design, it needs a relatively long balancing time, especially for the case that there are many other cells connected between the strongest and weakest cells in the battery pack. The equalization system with CPC topology shuttles energy among the cells and the pack. Since the cells are parts of the battery pack, it can be equivalent to that the cells can exchange energy with all other cells, which can provide faster equalizing speed than the cell-to-cell balancing system.

With utilizing the CPC equalization systems, many effective control algorithms have been designed for cell equalization of the battery pack. In [56], a current equalization method is designed based on each cell's capacity in the battery pack. This equalization method significantly contributes to increasing the discharge energy of the entire battery pack. In [1], an offline control protocol based on convex optimization is proposed for cells energy equalization. With a dual filter designed for state of balance estimation, a equalizing control algorithm based on the optimum reference is proposed in [57] for the high-power lithium-ion battery pack equalization. A batch-based run-to-run control strategy is proposed in [58] to control the cell equalizing current for active balancing of the battery pack to improve the capacity utilization, which can be implemented in a low-cost micro-controller unit.

Despite these achievements, there are still some improvements. Firstly, the conventional CPC equalization system is not suitable to be directly utilized for the large-scale battery packs, since the high voltage of the battery pack side brings a huge burden on the equalizer hardware. Moreover, the CPC structure brings challenges to the maintenance of the equalization system since it lacks of modular design. Due to these structural defects, it is hard to be well utilized in practical applications. Secondly, the cells' current constraints are not considered in the existing related work, which results that the equalizing current is designed independent of the battery pack's working current. It causes that the cells' currents may be larger than their maximum allowed bound when the battery pack works with a large charging or discharging current, thus bringing potential safety and health hazards to the battery pack.

Ouyang [59] aimed to achieve better equalization performance, dividing the battery pack into several modules and developing an improved module-based CPC equalization system, where the CMC/ML equalizers are utilized for the SOC equalization of the cells in each module/cell. It is more suitable for the battery pack consisting of numerous serially-connected cells than the conventional CPC balancing system due to its advantages of simple modular structure and convenient maintainability. Based on this improved equalization system, a two-layer MPC strategy is proposed with considering the current constraints, in which the ML equalizers are controlled by the top-layer MPC and the controlled CMC equalizing currents in each module are designed by the bottom-layer MPC algorithms in parallel. It can well guarantee the safety-related hard constraints of the cells' currents even when the battery pack works with a large external current. Moreover, its computational complexity is much less than the conventional centralized MPC, which makes it more feasible for real-time cell equalization implementation in practical applications.

2.3.3 Charging Equalization Algorithms

In addition to the balance between cells and the balance between cell and pack, charging equalization is also an important kind of balance methods. For charging control in battery packs, charging time, cell balance and temperature suppression are important factors to be considered. The most commonly utilized battery charging method is the CC-CV [60], in which a constant charging current is utilized until the battery's voltage reaches the specified value and then the mode is switched to a constant voltage charging. However, practical use of this method involves much empirical knowledge, e.g., deciding the magnitude of the current. Toward overcoming this issue, an optimal algorithm is proposed in [61] to determine the optimal constant current in the CC-CV method with consideration of charging time, energy loss, and temperature rise. Based on a coupled battery thermoelectric model, an advance optimal charging strategy is designed in [62] to develop the optimal CC-CV charging profile, which gives the best trade-off among the objectives of charging time, energy loss, and temperature rise. To improve the charging speed, a two-stage constant current charging scheme is formulated in [63], and an optimal five-step constant current charging strategy is proposed in [64]. In [65], an optimal voltage-based multistage constant current charging strategy is formulated and a multi-objective particle swarm optimization algorithm is employed to satisfy the different charging demands of charging time, charged capacity and energy loss with the impact of different weight combinations for balancing these demands analyzed.

Ouyang [66] designed a modified isolated buck converter based multi-module charger for a serially connected battery pack. Based on this charger, a hierarchical control approach is proposed to enable optimal multi-objective charging of a battery pack. This approach features a two-layer design. The top-layer control consists of three aspects: first, it received the user demand, which is expressed as the target SOC and charging time; second, it formulated a multi-objective optimization problem that comprehensively takes into account the user demand, cell equalization, temperature suppression, and operating constraints in the charging process; finally, it determined the optimal charging current profile by solving the optimization problem. In the bottom layer, an IDA-PBC is proposed to control the charging currents via the charger to track the current profile generated by the top-layer scheduling algorithm.

In addition, based on a multi-module charger [67], a user-involved strategy with the leader followers framework is proposed for the charging control of a series-connected lithium-ion battery pack. To be specific, by utilizing a nominal model of battery cells, an optimal average SOC trajectory is first generated through formulating and solving a multi-objective optimization issue with the consideration of both user demand and battery pack's energy loss. Next, a distributed charging strategy is proposed to make the cells' SOCs track the prescheduled trajectory, where observers are designed for online compensation of the cells' model bias. This strategy is that all cells track this trajectory simultaneously to achieve cell equalization, bringing the obvious benefits to significantly reduce computational effort in comparison with separately scheduling an optimal charging sequence for each cell.

To sum up, there are existing charging strategies all over the world. That only target single batteries with a single mode and lack of consideration for the consistency and balance of battery packs. The above two charging equalization algorithms propose a multi-objective charging control method for user-participatory battery packs, allowing users to participate in and set charging requirements, adjust the charging process under the premise of meeting safety constraints intelligently, and improve the intelligence while improving the battery pack capacity.

References

1. C. Pinto, J.V. Barreras, E. Schaltz, R.E. Araujo, Evaluation of advanced control for li-ion battery balancing systems using convex optimization. IEEE Trans. Sustain. Energy **7**(4), 1703–1717 (2016)
2. P.A. Cassani, S.S. Williamson, Design, testing, and validation of a simplified control scheme for a novel plug-in hybrid electric vehicle battery cell equalizer. IEEE Trans. Ind. Electron. **57**(12), 3956–3962 (2010)
3. S.R. Raman, X. Xue, K.E. Cheng, Review of charge equalization schemes for li-ion battery and super-capacitor energy storage systems, in *2014 International Conference on Advances in Electronics Computers and Communications*, Bangalore, India, pp. 1–6 (2014)
4. Y.S. Lee, M.W. Cheng, Intelligent control battery equalization for series connected lithium-ion battery strings. IEEE Trans. Ind. Electron. **52**(5), 1297–1307 (2005)
5. M. Daowd, N. Omar, P. Van Den Bossche, J. Van Mierlo, "Passive and active battery balancing comparison based on matlab simulation, in *IEEE Vehicle Power and Propulsion Conference*, Chicago, IL, USA, vol. 2011, pp. 1–7 (2011)
6. F. Altaf, B. Egardt, L. Johannesson Mårdh, Load management of modular battery using model predictive control: thermal and state-of-charge balancing. IEEE Trans. Control. Syst. Technol. **25**(1), 47–62 (2017)
7. Y. Liu, Research on power battery state observation and management system, Ph.D. dissertation, China University of Mining and Technology (2017)

8. C.S. Lim, K.J. Lee, N.J. Ku, D.S. Hyun, R.Y. Kim, A modularized equalization method based on magnetizing energy for a series-connected lithium-ion battery string. IEEE Trans. Power Electron. **29**(4), 1791–1799 (2014)
9. X. Lu, K. Sun, J.M. Guerrero, J.C. Vasquez, L. Huang, State-of-charge balance using adaptive droop control for distributed energy storage systems in DC microgrid applications. IEEE Trans. Ind. Electron. **61**(6), 2804–2815 (2014)
10. L.Y. Wang, C. Wang, G. Yin, F. Lin, M.P. Polis, C. Zhang, J. Jiang, Balanced control strategies for interconnected heterogeneous battery systems. IEEE Trans. Sustain. Energy **7**(1), 189–199 (2016)
11. M. Caspar, T. Eiler, S. Hohmann, Comparison of active battery balancing systems, in *IEEE Vehicle Power and Propulsion Conference (VPPC)*, Coimbra, Portugal, vol. 2014, pp. 1–8 (2014)
12. N. Nguyen, S.K. Oruganti, K. Na, F. Bien, An adaptive backward control battery equalization system for serially connected lithium-ion battery packs. IEEE Trans. Veh. Technol. **63**(8), 3651–3660 (2014)
13. S. Yarlagadda, T.T. Hartley, I. Husain, A battery management system using an active charge equalization technique based on a DC/DC converter topology. IEEE Trans. Ind. Appl. **49**(6), 2720–2729 (2013)
14. P.A. Cassani, S.S. Williamson, Feasibility analysis of a novel cell equalizer topology for plug-in hybrid electric vehicle energy-storage systems. IEEE Trans. Veh. Technol. **58**(8), 3938–3946 (2009)
15. N. Eghtedarpour, E. Farjah, Distributed charge/discharge control of energy storages in a renewable-energy-based DC micro-grid. IET Renew. Power Gener. **8**(1), 45–57 (2014)
16. D. Costinett, K. Hathaway, M.U. Rehman, M. Evzelman, R. Zane, Y. Levron, D. Maksimovic, Active balancing system for electric vehicles with incorporated low voltage bus, in *IEEE Applied Power Electronics Conference and Exposition-APEC 2014*, Fort Worth, TX, USA, vol. 2014, pp. 3230–3236 (2014)
17. R. Ling, Q. Dan, L. Wang, D. Li, Energy bus-based equalization scheme with bi-directional isolated CUK equalizer for series connected battery strings, in *IEEE Applied Power Electronics Conference and Exposition (APEC)*, Charlotte, NC, USA, vol. 2015, pp. 3335–3340 (2015)
18. M.M.U. Rehman, M. Evzelman, K. Hathaway, R. Zane, G.L. Plett, K. Smith, E. Wood, D. Maksimovic, Modular approach for continuous cell-level balancing to improve performance of large battery packs, in *IEEE energy conversion congress and exposition (ECCE)*, Pittsburgh, PA, USA, vol. 2014, pp. 4327–4334 (2014)
19. Y. Wang, C. Zhang, Z. Chen, J. Xie, X. Zhang, A novel active equalization method for lithium-ion batteries in electric vehicles. Appl. Energy **145**, 36–42 (2015)
20. Y. Shang, Q. Zhang, N. Cui, C. Zhang, A cell-to-cell equalizer based on three-resonant-state switched-capacitor converters for series-connected battery strings. Energies **10**(2), 1–15 (2017)
21. T.M. Bui, C-.H. Kim, K-.H. Kim, S.B. Rhee, A modular cell balancer based on multi-winding transformer and switched-capacitor circuits for a series-connected battery string in electric vehicles. Appl. Sci. **8**(8), 1278 (2018). https://doi.org/10.3390/app8081278
22. Y. Chen, X. Liu, H.K. Fathy, J. Zou, S. Yang, A graph-theoretic framework for analyzing the speeds and efficiencies of battery pack equalization circuits. Int. J. Electr. Power Energy Syst. **98**, 85–99 (2018)
23. Y. Shang, C. Zhang, N. Cui, C.C. Mi, A delta-structured switched-capacitor equalizer for series-connected battery strings. IEEE Trans. Power Electron. **34**(1), 452–461 (2019)
24. Y. Shang, N. Cui, B. Duan, C. Zhang, Analysis and optimization of star-structured switched-capacitor equalizers for series-connected battery strings. IEEE Trans. Power Electron. **33**(11), 9631–9646 (2018)
25. A.C. Baughman, M. Ferdowsi, Double-tiered switched-capacitor battery charge equalization technique. IEEE Trans. Ind. Electron. **55**(6), 2277–2285 (2008)
26. T.H. Phung, A. Collet, J.C. Crebier, An optimized topology for next-to-next balancing of series-connected lithium-ion cells. IEEE Trans. Power Electron. **29**(9), 4603–4613 (2013)

27. K. Friansa, I.N. Haq, E. Leksono, N. Tapran, D. Kurniadi, B. Yuliarto, Battery module performance improvement using active cell balancing system based on switched-capacitor boost converter (s-cbc), in *2017 4th International Conference on Electric Vehicular Technology (ICEVT)*, Indonesia, pp. 93–99 (2017)
28. X. Wang, K.W.E. Cheng, Y.C. Fong, Non-equal voltage cell balancing for battery and supercapacitor source package management system using tapped inductor techniques. Energies **11**(5), 1037 (2018)
29. X. Liu, Z. Wan, Y. He, X. Zheng, G. Zeng, J. Zhang, A unified control strategy for inductor-based active battery equalisation schemes. Energies **11**(2), 405 (2018)
30. A.F. Moghaddam, A. Van Den Bossche, An active cell equalization technique for lithium ion batteries based on inductor balancing, in *2018 9th International Conference on Mechanical and Aerospace Engineering (ICMAE)*, Budapest, Hungary, pp. 274–278 (2018)
31. A. Farzan Moghaddam, A. Van den Bossche, An efficient equalizing method for lithium-ion batteries based on coupled inductor balancing. Electronics **8**(2), 1–12 (2019)
32. C. Zhang, Y. Shang, Z. Li, N. Cui, An interleaved equalization architecture with self-learning fuzzy logic control for series-connected battery strings. IEEE Trans. Veh. Technol. **66**(12), 10 923–10 934 (2017)
33. X. Liu, Y. Sun, Y. He, X. Zheng, G. Zeng, J. Zhang, Battery equalization by fly-back transformers with inductance, capacitance and diode absorbing circuits. Energies **10**(10), 1482 (2017). https://doi.org/10.3390/en10101482
34. Y. Ye, K.W.E. Cheng, Analysis and design of zero-current switching switched-capacitor cell balancing circuit for series-connected battery/supercapacitor. IEEE Trans. Veh. Technol. **67**(2), 948–955 (2018)
35. M. Liu, M. Fu, Y. Wang, C. Ma, Battery cell equalization via megahertz multiple-receiver wireless power transfer. IEEE Trans. Power Electron. **33**(5), 4135–4144 (2018)
36. S. Li, C.C. Mi, M. Zhang, A high-efficiency active battery-balancing circuit using multiwinding transformer. IEEE Trans. Ind. Appl. **49**(1), 198–207 (2013)
37. Y. Chen, X. Liu, Y. Cui, J. Zou, S. Yang, A multiwinding transformer cell-to-cell active equalization method for lithium-ion batteries with reduced number of driving circuits. IEEE Trans. Power Electron. **31**(7), 4916–4929 (2015)
38. S. Li, C. Mi, M. Zhang, A high efficiency low cost direct battery balancing circuit using a multi-winding transformer with reduced switch count, in *Twenty-Seventh Annual IEEE Applied Power Electronics Conference and Exposition (APEC)*, Orlando, USA, vol. 2012, pp. 2128–2133 (2012)
39. X. Ding, D. Zhang, J. Cheng, B. Wang, Y. Chai, Z. Zhao, R. Xiong, P.C.K. Luk, A novel active equalization topology for series-connected lithium-ion battery packs. IEEE Trans. Ind. Appl. **56**(6), 6892–6903 (2020)
40. A.F. Moghaddam, A. Van den Bossche, A battery equalization technique based on Ćuk converter balancing for lithium ion batteries, in *2019 8th International Conference on Modern Circuits and Systems Technologies (MOCAST)*, Thessaloniki, Greece, pp. 1–4 (2019)
41. J. Gallardo-Lozano, E. Romero-Cadaval, M.I. Milanes-Montero, M.A. Guerrero-Martinez, Battery equalization active methods. J. Power Sources **246**, 934–949 (2014)
42. S. Maniktala, *Switching Power Supplies A-Z* (Elsevier, 2012)
43. D. Maksimovic, S. Cuk, A unified analysis of PWM converters in discontinuous modes. IEEE Trans. Power Electron. **6**(3), 476–490 (1991)
44. L. Yuang-Shung, D. Jiun-Yi, Fuzzy-controlled individual-cell equaliser using discontinuous inductor current-mode Cuk convertor for lithium-ion chemistries. IEE Proc.-Electr. Power Appl. **152**(5), 1271–1282 (2005)
45. Y.S. Lee, D. Cheng, S. Wong, A new approach to the modeling of converters for spice simulation. IEEE Trans. Power Electron. **7**(4), 741–753 (1992)
46. J. Yan, Z. Cheng, G. Xu, H. Qian, Y. Xu, Fuzzy control for battery equalization based on state of charge, in *IEEE 72nd Vehicular Technology Conference-Fall*, Ottawa, ON, Canada, vol. 2010, pp. 1–7 (2010)

47. H. Li, Interpolation mechanism of fuzzy control. Sci. China Ser. E: Technol. Sci. **41**(3), 312–320 (1998)
48. J. Zheng, J. Chen, Q. Ouyang, Variable universe fuzzy control for battery equalization. J. Syst. Sci. Complex. **31**(1), 325–342 (2018)
49. M.F. Samadi, M. Saif, Nonlinear model predictive control for cell balancing in li-ion battery packs, in *American Control Conference*, Portland, OR, USA, vol. 2014, pp. 2924–2929 (2014)
50. Q. Ouyang, J. Chen, J. Zheng, Y. Hong, SOC estimation-based quasi-sliding mode control for cell balancing in lithium-ion battery packs. IEEE Trans. Ind. Electron. **65**(4), 3427–3436 (2018)
51. S. Abhinav, G. Binetti, A. Davoudi, F.L. Lewis, Toward consensus-based balancing of smart batteries, in *IEEE Applied Power Electronics Conference and Exposition-APEC 2014*, Fort Worth, TX, USA, vol. 2014, pp. 2867–2873 (2014)
52. R. Ling, Q. Dan, J. Zhang, G. Chen, A distributed equalization control approach for series connected battery strings, in *The 26th Chinese Control and Decision Conference, (CCDC)*, Changsha, China, vol. 2014, pp. 5102–5106 (2014)
53. C. Li, E.A.A. Coelho, T. Dragicevic, J.M. Guerrero, J.C. Vasquez, Multiagent-based distributed state of charge balancing control for distributed energy storage units in AC microgrids. IEEE Trans. Ind. Appl. **53**(3), 2369–2381 (2017)
54. Q. Ouyang, J. Chen, J. Zheng, H. Fang, Optimal cell-to-cell balancing topology design for serially connected lithium-ion battery packs. IEEE Trans. Sustain. Energy **9**(1), 350–360 (2018)
55. Q. Ouyang, W. Han, C. Zou, G. Xu, Z. Wang, Cell balancing control for lithium-ion battery packs: a hierarchical optimal approach. IEEE Trans. Ind. Inform. **16**(8), 5065–5075 (2020)
56. M. Einhorn, W. Guertlschmid, T. Blochberger, R. Kumpusch, R. Permann, F.V. Conte, C. Kral, J. Fleig, A current equalization method for serially connected battery cells using a single power converter for each cell. IEEE Trans. Veh. Technol. **60**(9), 4227–4237 (2011)
57. S. Wang, L. Shang, Z. Li, H. Deng, J. Li, Online dynamic equalization adjustment of high-power lithium-ion battery packs based on the state of balance estimation. Appl. Energy **166**, 44–58 (2016)
58. X. Tang, C. Zou, T. Wik, K. Yao, Y. Xia, Y. Wang, D. Yang, F. Gao, Run-to-run control for active balancing of lithium iron phosphate battery packs. IEEE Trans. Power Electron. **35**(2), 1499–1512 (2020)
59. Q. Ouyang, Y. Zhang, N. Ghaeminezhad, J. Chen, Z. Wang, X. Hu, J. Li, Module-based active equalization for battery packs: a two-layer model predictive control strategy. IEEE Trans. Transp. Electrification **8**(1), 149–159 (2022)
60. S.S. Zhang, K. Xu, T. Jow, Study of the charging process of a licoo2-based li-ion battery. J. Power Sources **160**(2), 1349–1354 (2006)
61. A. Abdollahi, X. Han, G. Avvari, N. Raghunathan, B. Balasingam, K.R. Pattipati, Y. Bar-Shalom, Optimal battery charging, part i: minimizing time-to-charge, energy loss, and temperature rise for OCV-resistance battery model. J. Power Sources **303**, 388–398 (2016)
62. K. Liu, K. Li, Z. Yang, C. Zhang, J. Deng, An advanced lithium-ion battery optimal charging strategy based on a coupled thermoelectric model. Electrochimica Acta **225**, 330–344 (2017)
63. S.S. Zhang, The effect of the charging protocol on the cycle life of a li-ion battery. J. power sources **161**(2), 1385–1391 (2006)
64. Y.H. Liu, C.H. Hsieh, Y.F. Luo, Search for an optimal five-step charging pattern for li-ion batteries using consecutive orthogonal arrays. IEEE Trans. Energy Convers. **26**(2), 654–661 (2011)
65. H. Min, W. Sun, X. Li, D. Guo, Y. Yu, T. Zhu, Z. Zhao, Research on the optimal charging strategy for li-ion batteries based on multi-objective optimization. Energies **10**(5), 1–15 (2017)
66. Q. Ouyang, J. Chen, J. Zheng, H. Fang, Optimal multiobjective charging for lithium-ion battery packs: a hierarchical control approach. IEEE Trans. Ind. Inform. **14**(9), 4243–4253 (2018)
67. Q. Ouyang, Z. Wang, K. Liu, G. Xu, Y. Li, Optimal charging control for lithium-ion battery packs: a distributed average tracking approach. IEEE Trans. Ind. Inf. **16**(5), 3430–3438 (2020)

Chapter 3
Active Cell Equalization Topology

3.1 Commonly Used Active Cell Equalization Topology

Most topologies in battery balancing systems are concerned with the system-level perspective of cells and equalizers. For a better understanding, consider the topologies as graphs in which the cells are nodes and each cell equalizer is regarded as an edge that acts as an energy transfer link between the cells (nodes) [1, 2]. The active cell equalization systems responsible for transferring energy flow can be classified into four types, as shown in Fig. 3.1: adjacent-based, non-adjacent-based, direct cell-cell, and mixed topologies. These topologies would be further selected depending on the structures listed below.

- Adjacent-based topology makes use of active equalizers that are performed on each adjacent cell or module of a battery pack to measure the SOC or average SOC level and balance it across the entire battery pack using a specific energy transferring approach from the high cell or module SOC level to the lower one [3, 4]. These balancing topologies are widely used in the study of battery balancing applications [5, 6], and could be further classified as series-based, module-based, and layer-based cell-cell topologies [7]. Figure 2.2a, b show two active cell equalization circuits with the cell-cell topology.
- Non-adjacent-based topology uses active equalizers in each cell or module and the entire battery pack to measure and balance the SOC levels of the cells or modules based on a defined energy flow strategy that transfers charge between cells or modules and the pack [7, 8]. Subsequently, the non-adjacent-based topology is divided into series-based and module-based cell-pack in this paper. For illustration, Fig. 2.2c, d, represent two active cell equalization circuits with the cell-pack topology.
- Direct cell-cell topology is used to effectively transmit energy from one cell to another cell utilizing a standard equalizer with a couple of switches with each cell as well as other electrical components such as a capacitor, irrespective of whether the cells are contiguous or non-contiguous. To avoid a short circuit, only the switches associated with the cell whose SOC has the greatest bias with the

J. Chen et al., *Equalization Control for Lithium-Ion Batteries*,
https://doi.org/10.1007/978-981-99-0220-0_3

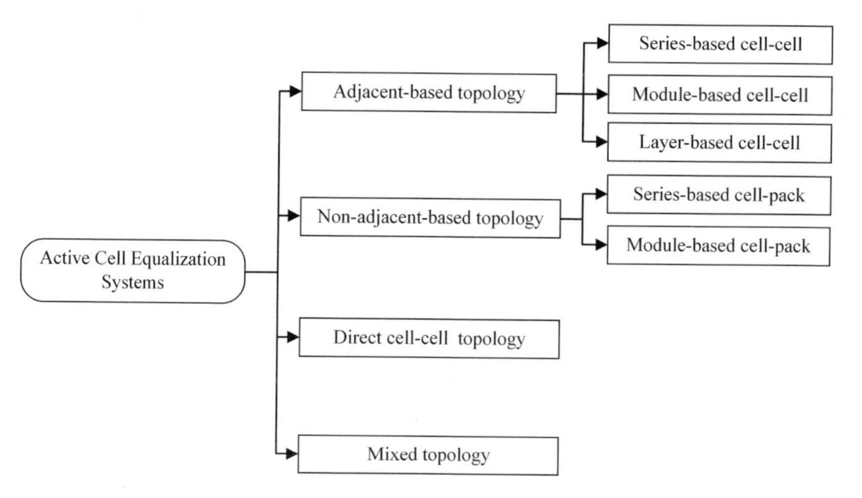

Fig. 3.1 Classification of active equalization systems

pack's average SOC can indeed be closed for every typical working time. As a result of consistently connecting the switches connected with all those individual cells, the SOC level of the entire pack achieves the same level [9–11].

- The mixed topology has two interrelated balancing layers that use either adjacent-based or non-adjacent-based topologies in each layer. Following that, the battery pack is partitioned into several components, each containing the same quantity of cells. The bottom layer balances the cells inside every module using the attributed topology, which is controlled by a slave BMS. Following that, the top layer has the responsibility of having to balance the charge between the modules via the associated topology, which is controlled by a master BMS. The slave and master BMS are linked, and it appears that this topology could provide incorporation of both top and bottom layer topology advantages [12].

3.1.1 Adjacent-Based Topology

The energy can be transferred solely between adjacent modules or cells of a battery pack for the balancing systems having adjacent-based topology. However, in a battery pack, the module with higher energy could manage to deliver the surplus amount of energy to the module or adjacent cell of the same battery pack. Furthermore, the module or cell with lower energy is warranted to recover the demanded amount of energy from adjacent module or cell to maintain the SOC level of all the associated cells in the battery pack. The adjacent-based topologies are classified into three types based on their topology-level structure: series-based, module-based, and layer-based [7].

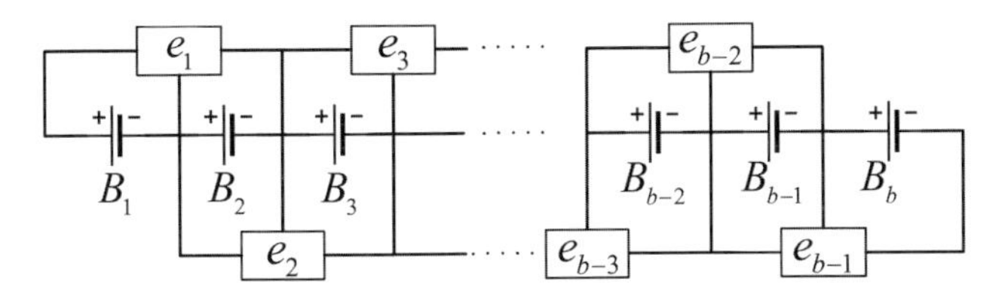

Fig. 3.2 Series-based cell-cell battery charge equalization topology [7]

Series-based cell-cell Topology: The series-based topological structure, as depicted in Fig. 3.2 [7, 13], is the most widely used of all the previously defined adjacent-based topological structures. Accordingly, in every charging/discharging time slots of a battery pack with b number of serially connected battery cells, the equalizer $e_i (0 \leq i \leq b-1)$, transfers $\epsilon_i(k)$ amount of energy from B_i to $B_{i+1} (1 \leq i \leq b-1)$ battery cells and vice versa, considering the cell with the lowest SOC level to be the destination equalizing cell. Therefore, there is no energy transfer between two adjacent cells that have the same SOC level [14, 15]. The mathematical model of cell equalizing system's with series-based topological structure for a battery pack consisting of b serially connected cells can be calculated as [3]:

$$x_i(k+1) = x_i(k) + d(\epsilon_{i-1}(k) - \epsilon_i(k)), i = 1, ..., b \tag{3.1}$$

with $d = \frac{\eta_0}{Q}$, where the parameters η_0 and Q are the Columbic efficiency and the cells' nominal capacity; $x_i(k)$ defines the SOC level of the cell B_i. It is to be noted that $\epsilon_0(k) = \epsilon_b(k) = 0$, and $\epsilon_i(k)(1 \leq i \leq b-1)$ can be represented as follows:

$$\epsilon_i(k) = (k_i + k_i')u_i(k)T \tag{3.2}$$

where

$$\begin{cases} k_i = 1, k_i' = 0 & \text{if } x_i(k) > x_{i+1}(k) \\ k_i = 0, k_i' = 0 & \text{if } x_i(k) = x_{i+1}(k) \\ k_i = 0, k_i' = -1 & \text{if } x_i(k) < x_{i+1}(k) \end{cases}$$

with $u_i(k) \geq 0$ is the equalizing current to be controlled through the i-th equalizer. Also, note that $u_i(k)$ is the equalizing current from the i-th [(i+1)-th] cell to the (i+1)-th (i-th) cell when $x_i > x_{i+1}$ ($x_{i+1} > x_i$), and T is time period of the control sample.

A large number of algorithms have been proposed in the literature for designing the controlled equalizing current by incorporating a series-based topological structure. For example, [17] incorporates an FLC based algorithm for balancing the charge between series connected battery cells and significantly reduces the equalizing time. The authors of [18] achieved voltage level equalization for neighboring battery cells connected in series by proposing a nonlinear dynamic control technique integrating neural networks and a fuzzy logic based control algorithm. Following that, a neuron-

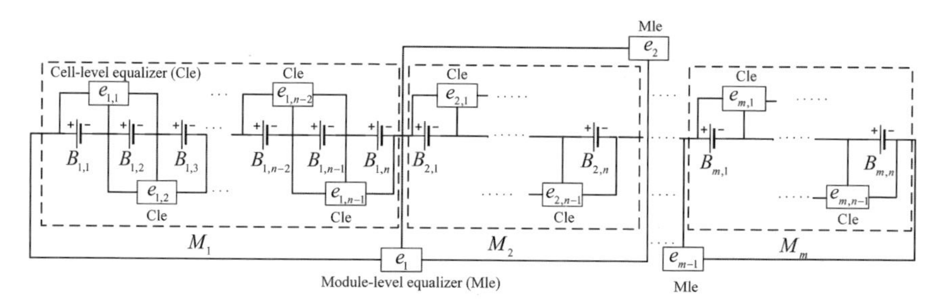

Fig. 3.3 Module-based cell-cell battery charge equalization topology [16]

fuzzy controller was incorporated to tune the duty cycle of the switches utilized in the equalization circuit, advancing the robustness of the equalization process and reducing equalizing time. In addition to the previous discussed examples, this kind of topological structure can be applied to the nonlinear MPCs, as propounded in [19] for balancing the SOC level of the serially connected cells within a battery pack. In reference to this type of model, the total power loss, the equalizing time, and the difference between SOCs of the adjacent cells are minimized.

Module-based cell-cell topology: This type of topological drive operates similarly to a series-based topological structure in terms of charging and discharging, but takes adjacent modules into account for equalization. Fitting for this purpose, each battery pack with the total $b = mn$ battery cells, is divided into m modules $M_1, M_2, \ldots, M_m$, which individually contains equal n cells, as shown in Fig. 3.3 [4, 7, 13]. In each module, there are $n - 1$ Cle, $e_{j,i}(1 \leq j \leq m, 1 \leq i \leq n - 1)$, which balance the SOC level between adjacent cells by transferring $\epsilon_{j,i}(k)$ amount of energy from higher SOC level cells to the lower one. Furthermore, taking this to the module level, $m - 1$ number of Mle are exploited to balance the SOC level of all adjacent modules. To accomplish this, the modules $M_j(1 \leq j \leq m)$ in the battery pack are thought of in such a way that the equalizer $e_j(1 \leq j \leq m - 1)$ releases $\epsilon_j(k)$ quantity of energy from every single cell of the module with higher average SOC level to the lower one in the same proportion. There is obviously no charge transferring on an equal SOC average between two adjacent modules [20, 21]. Because of their serial connection, all of the cells within each battery module receive the same module-level equalizing current. A module-based topological structure's cell equalizing mathematical model for a battery pack are made up of $b = mn$ serially connected cells can be calculated as [16]:

$$x_{j,i}(k+1) = x_{j,i}(k) + d(\epsilon_{j-1}(k) - \epsilon_j(k) + \epsilon_{j,i-1}(k) - \epsilon_{j,i}(k)) \tag{3.3}$$

where $x_{j,i}(k)$ is the i-th $(1 \leq i \leq n)$ cells SOC of the corresponding j-th $(1 \leq j \leq m)$ battery module; $\epsilon_0(k) = \epsilon_m(k) = \epsilon_{j,0}(k) = \epsilon_{j,n}(k) = 0$; $\epsilon_j(k)$ and $\epsilon_{j,i}(k)$ can be calculated as:

$$\begin{aligned} \epsilon_j(k) &= (k_j + k'_j)\frac{u_j(k)}{n}T \\ \epsilon_{j,i}(k) &= (k_{j,i} + k'_{j,i})u_{j,i}(k)T \end{aligned} \tag{3.4}$$

with

$$\begin{cases} k_j = 1, k'_j = 0 & \text{if } \bar{x}_j(k) > \bar{x}_{j+1}(k) \\ k_j = 0, k'_j = 0 & \text{if } \bar{x}_j(k) = \bar{x}_{j+1}(k) \\ k_j = 0, k'_j = -1 & \text{if } \bar{x}_j(k) < \bar{x}_{j+1}(k) \end{cases}$$

$$\begin{cases} k_{j,i} = 1, k'_{j,i} = 0 & \text{if } x_{j,i}(k) > x_{j,i+1}(k) \\ k_{j,i} = 0, k'_{j,i} = 0 & \text{if } x_{j,i}(k) = x_{j,i+1}(k) \\ k_{j,i} = 0, k'_{j,i} = -1 & \text{if } x_{j,i}(k) < x_{j,i+1}(k) \end{cases}$$

where $\bar{x}_j(k) = \frac{1}{n}\sum_{i=1}^{n} x_{j,i}(k)$, in which $u_j(k)$ and $u_{j,i}(k)$ are the controlled equalizing current through the j-th Mle and i-th Cle of j-th module, respectively; $\bar{x}_j(k)$ is the average SOC level of the j-th module.

Different control structures can be incorporated to optimize and accomplish the SOC equalization process in cells. The authors have incorporated a re-configurable cell/module algorithm in [6], in which the propounded battery charge equalization system is made up of series-based equalizers acting on adjacent cells within which the module-based equalizers are positioned. Consequently, the charge balance performance has improved and been optimized. Furthermore, when compared to exhaustive search and genetic algorithms, the battery charge equalizing time can be minimized by the propounded algorithms while equalizing at a faster or near-faster rate. Similar to [6], the authors in [4] have incorporated the same module-based battery charging equalization system. However, for measuring the critical performance and transitory cells SOC, the authors proposed some estimation techniques. By incorporating this technique, the computational efficiency is improved along with accurate estimation in the performed numerical experiments. On the other hand, a hierarchical control algorithm for cell equalizing with multiple objectives are propounded in [16]. Following that, the proposed technique converges the cells SOC level in the battery pack quickly while restricting SOCs, equalizing currents, and currents.

Layer-based cell-cell Topology: The equalization topology does have a binary-tree based structuring with in course of layer-based topological structure, as shown in Fig. 3.4. Accordingly, with b serially connected cells in a battery pack $B_i (1 \leq i \leq b)$, there can be the total number of $\log_2{}^b$ layers, in which the layer $l(1 \leq l \leq \log_2{}^b)$ contains $b/2^l$ equalizers $e_{l,j}(1 \leq j \leq b/2^l)$. Following that, two groups $G^-_{l,j}$ and $G^+_{l,j}$ of battery cells are used to represent all the connected cells with equalizer $e_{l,j}$, where $G^-_{l,j} = \{(j1)2^l + 1, (j1)2^l + 2, \ldots, (j1)2^l + 2^{l1}\}$, and $G^+_{l,j} = \{(2j1)2^{l-1} + 1, (2j1)2^{l-1} + 2, \ldots, (2j1)2^{l-1} + 2^{l1}\}$. The equalizer $e_{l,j}$ transfers excess energy from each battery cell throughout the battery cohort with the higher overall SOC to the battery cohort with the lower overall SOC. Consider the fact that, as a result of their serial connection, all cells share the same layer level equalizing current within each battery layer. In the first layer, each pair of adjacent cells is connected by an

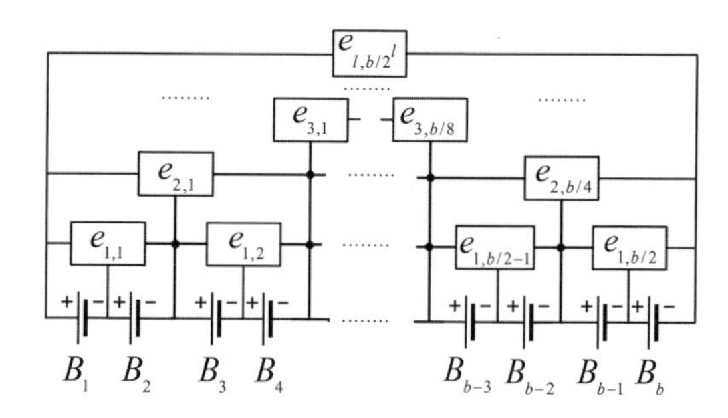

Fig. 3.4 Layer-based cell-cell battery charge equalization topology [23]

equalizer, as shown in Fig. 3.4. Contemporaneously, the topmost layer equalizers stabilize the SOC level of the battery cells in relation to the two adjacent lower layer equalizers. Simultaneously, the energy can be transferred similar to the module-based topological structure [7, 22–24].

The cell equalizing system dynamics of a layer-based cell-to-cell topological structure for a battery pack comprising of b series—connected cells could be captured by the following mathematical model:

$$x_i(k+1) = x_i(k) + d \sum_{l=1}^{\lceil \log_2{}^b \rceil} (-1)^{\lfloor \frac{i}{2^l} \rfloor} \epsilon_{l,\lceil \frac{i}{2^l} \rceil}(k) \tag{3.5}$$

where $x_i(k)$ indicates the i-th $(1 \le i \le b)$ cells SOC; $\lceil \cdot \rceil$ and $\lfloor \cdot \rfloor$ are the ceiling and floor functions that round up and down the $\cdot$ to the nearby integer, respectively; $\epsilon_{l,j}(k)$ $(1 \le l \le \log_2{}^b,\ 1 \le j \le b/2^l)$ is:

$$\epsilon_{l,j}(k) = (k_{l,j} + k'_{l,j}) \frac{u_{l,j}(k)}{2^{l-1}} T \tag{3.6}$$

with

$$\begin{cases} k_{l,j} = 1, k'_{l,j} = 0 & \text{if } \sum_{g \in G^-_{l,j}} x_g(k) > \sum_{g \in G^+_{l,j}} x_g(k) \\ k_{l,j} = 0, k'_{l,j} = 0 & \text{if } \sum_{g \in G^-_{l,j}} x_g(k) = \sum_{g \in G^+_{l,j}} x_g(k) \\ k_{l,j} = 0, k'_{l,j} = -1 & \text{if } \sum_{g \in G^-_{l,j}} x_g(k) < \sum_{g \in G^+_{l,j}} x_g(k) \end{cases}$$

with $u_{l,j}(k)$ defines the controlled equalizing current flowing through the j-th equalizer of l-th layer; $x_g(k)$ indicates the SOC level of the cell g ($g \in G^-_{l,j}$or $g \in G^+_{l,j}$).

The layer-based architecture for battery charge equalization is shown as an example in [22]. According to the propounded architecture, the equalizers are separately placed in various layers and start behaving on the respective battery cells concurrently, significantly lowering the battery charge equalization time and thus improving the energy storage system performance. The layer-based topological structure is incorporated in the battery charge equalization in another research [23]. A decentralized topological structure is propounded to achieve the cell equalizing function. The equalizing nodes operate independently and can be linked together to form a multi-layer distributed equalizing structure, allowing the entire system to balance the additional total connected battery cells. As a result of the proposed exclusivity structure, the modular design becomes more flexible and straightforward, while also increasing battery capacity and prolonging cycle life. In [25], the authors propounded a hierarchical topological structure for simultaneous equalization to operate within and between cohorts of the series-connected battery cells by incorporating a fast control algorithm. In consequence, the simulation results and experimental implementation demonstrate that the propounded battery equalization system has achieved quicker and more efficient battery pack equalization than the equilibrium strategic approach, which only equalizes the first and subsequent groups of battery cells.

3.1.2 Non-adjacent-Based Topology

The non-adjacent-based topology is a type of active equalization technique that achieves cell equalization by commuting extra energy throughout a battery pack. Equalization topologies for the charge transferring arrangement of cells within a battery pack are classified into three types: pack-to-cell, cell-to-pack, and cell-to-pack-to-cell. The associated equalizer, in particular for the cell-to-pack topological structure, delivers the cell's excess charge with the highest SOC level to the entire battery pack [11, 25–27]. The charge is transferred from the entire battery pack to an individual cell in the case of the pack-to-cell topology. Nonetheless, the equalizer transfers charge from the set cells to the entire pack while continuing to the desired cells throughout the cell-to-pack-to-cell topology [26, 28–31]. As a result, the proposed cell-pack topology here only includes the cell-to-pack-to-cell topology, which combines the working styles of cell-to-pack and pack-to-cell and has significantly better performance [32, 33].

In accordance with the topology-level classification in this section, the non-adjacent-based topology is classified as series-based and module-based cell-pack.

Series-based cell-pack topology: A series-based cell-pack topological structure can equalize b number of series-connected cells in a battery pack $B_i(1 \leq i \leq b)$, as shown in Fig. 3.5, where the pack equalizer e_p charges or discharges each cell individually in parallel with other cells. In consequence, $\epsilon_{i_p}(k)$ energy will be transferred from the pack to the i-th battery cell if its SOC is less than the average SOC of the pack. Otherwise, the battery pack receives $\epsilon_{i_p}(k)$ amount of energy from each cell simultaneously. This energy flow model provides a battery pack that may be fully charged or discharged without risk of overcharging or overdischarging the cells [28–31, 34–36]. The following expression can be used to calculate the model of the cell equalizing system in a series-based cell-pack topological structure for a battery pack made up of b series—connected cells [37]:

$$x_i(k+1) = x_i(k) + d\epsilon_{i_p}(k) \tag{3.7}$$

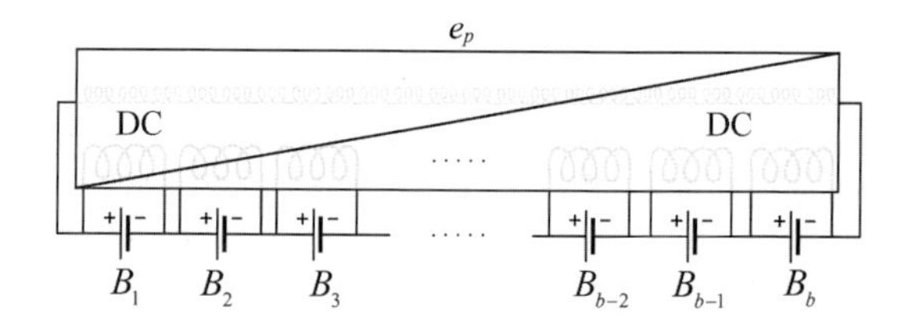

Fig. 3.5 Series-based cell-pack battery charge equalization topology [37]

where $x_i(k)(1 \leq i \leq b)$ is the SOC level of the battery cell B_i; $\epsilon_{i_p}(k)(1 \leq i \leq b)$ can be expressed as follows:

$$\epsilon_{i_p}(k) = \left(k_{i_p}\frac{1}{b}\sum_{i=1}^{b} u_i(k) + k'_{i_p}u_i(k)\right)T \tag{3.8}$$

with

$$\begin{cases} k_{i_p} = 0, k'_{i_p} = 1 & \text{if } \bar{x}(k) > x_i(k) \\ k_{i_p} = 0, k'_{i_p} = 0 & \text{if } \bar{x}(k) = x_i(k) \\ k_{i_p} = -1, k'_{i_p} = 0 & \text{if } \bar{x}(k) < x_i(k) \end{cases}$$

where $\bar{x}(k) = \frac{1}{b}\sum_{i=1}^{b} x_i(k)$, in which $u_i(k)$ is the controlled equalizing current through the pack equalizer; $\bar{x}(k)$ defines the average SOC level of the whole battery pack.

The effectiveness and overall performance of the series-based cell-pack topological structure have been demonstrated in [29], in which the series-based cell-pack topology has been set off for a charge equalization technique, which is conductive to serially-connected battery cells with varying cell capacities. In this method, the SOC of each battery cell remains directly proportional to its capacity while charging and discharging; thus, the whole battery pack behaves similarly to a single cell with a higher voltage. The proposed equalizing method in [31], which uses an energy coupling transformer as the equalizer, is another example of battery cell equalization found on series-based cell-pack topology. In accordance with low to high power applications, the equalization technique utilizes a simple yet effective topological structure that incorporates various types of switching converters. In [38], the authors have proposed a novel cell voltage equalizer that is based on a time-shared flyback converter controlled by a microcontroller and shared between cells. The major advantages of this type of equalizer are the significant cost reduction and improved efficiency. The optimum equalizing efficiency has been obtained with minimum power path by incorporating a flyback equalizer in [39] for selecting the target cell and performing the charge equalization process. Subsequently, for accelerating the equalization speed, the peak current mode control strategy can be incorporated.

Module-based cell-pack topology: This type of topological structure is shown in Fig. 3.6. We can see that there are b serially connected cells in a battery pack with $B_i(1 \leq i \leq b)$ is partitioned into m modules, $M_1, M_2, ..., M_m$, and every

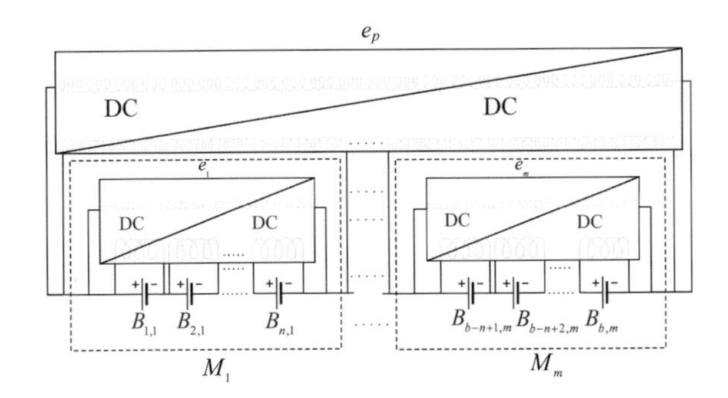

Fig. 3.6 Module-based cell-pack battery charge equalization topology

module containing equal number of battery cells n. In accordance, every module $M_j(1 \leq j \leq m)$ incorporates a single equalizer similar to the one used in the series-based topological structure, e_j. This transfers $\epsilon_{(j,i)_p}(k)$ energy from the respective module pack to cells with SOCs less than the module's average SOC, or from every cell of the designated module to that module pack. A central equalizer e_p has been located in the top layer, having the same operating principle as the bottom layer, but within the average modular SOC levels. At this moment, $\epsilon_{j_p}(k)$ energy can be delivered to every cells of M_j by the equalizer e_p, if the average SOC level of the module M_j is less than the average SOC of all modules. Apart from that, the central equalizer e_p in top layer receives $\epsilon_{j_p}(k)$ energy from each cell of all modules. The module-based topological structure is less complex and more efficient than the previously discussed series-based topological structure, which has the shortcoming of a very complex control algorithm, which limits the cell extension in a battery pack. Indeed, when a large number of battery cells are used for equalization, this topological structure has a significant advantage over series-based topological structures in terms of significantly lower energy loss and more distinct equalization power [26, 32, 33, 40]. The mathematical model of a module-based cell equalization topology for a battery pack with $b = mn$ serially connected cells can be formulated as:

$$x_{j,i}(k+1) = x_{j,i}(k) + d(\epsilon_{j_p}(k) + \epsilon_{(j,i)_p}(k)) \tag{3.9}$$

where $x_{j,i}(k)$ defines the SOC level of the i-th $(1 \leq i \leq n)$ cell in the j-th $(1 \leq j \leq m)$ module; $\epsilon_{j_p}(k)$ and $\epsilon_{(j,i)_p}(k)$ can be obtained as:

$$\begin{aligned} \epsilon_{j_p}(k) &= (k_{j_p} \tfrac{1}{b} \sum_{j=1}^{m} u_j(k) + k'_{j_p} \tfrac{u_j(k)}{n})T \\ \epsilon_{(j,i)_p}(k) &= (k_{(j,i)_p} \tfrac{1}{n} \sum_{i=1}^{n} u_{j,i}(k) + k'_{(j,i)_p} u_{j,i}(k))T \end{aligned} \tag{3.10}$$

with

$$\begin{cases} k_{j_p} = 0, k'_{j_p} = 1 & \text{if } \bar{x}(k) > \bar{x}_j(k) \\ k_{j_p} = 0, k'_{j_p} = 0 & \text{if } \bar{x}(k) = \bar{x}_j(k) \\ k_{j_p} = -1, k'_{j_p} = 0 & \text{if } \bar{x}(k) < \bar{x}_j(k) \end{cases}$$

$$\begin{cases} k_{(j,i)_p} = 0, k'_{(j,i)_p} = 1 & \text{if } \bar{x}_j(k) > x_{j,i}(k) \\ k_{(j,i)_p} = 0, k'_{(j,i)_p} = 0 & \text{if } \bar{x}_j(k) = x_{j,i}(k) \\ k_{(j,i)_p} = -1, k'_{(j,i)_p} = 0 & \text{if } \bar{x}_j(k) < x_{j,i}(k) \end{cases}$$

where $\bar{x}(k) = \frac{1}{m}\sum_{j=1}^{m} \bar{x}_j(k)$, in which $u_j(k)$, and $u_{j,i}(k)$ are the controlled equalizing current through the pack equalizer e_p, and module pack equalizer $e_j (1 \leq j \leq m)$, respectively; $\bar{x}_j(k)$ is defined in (3.4), and $\bar{x}(k)$ counts for the average SOC level of the whole modules.

An application of module-based cell-pack topology has been done in [26], in which, the authors have proposed a novel design of module charge equalizer which utilizes a multi-winding transformer. Equalization of the modules is accomplished by utilising the force of magnetizing energy. In consequence, the propounded equalizing structure restricts the size, cost, and increase of loss caused by modularization. In [32], the authors employ a module-based topological structure to accomplish cell equalization in a serially-connected battery pack. For changing the charging current direction, the authors have incorporated a two-stage DC-DC converter which also reduces the complexity and volume of the equalization circuits. In [33], another illustration of module-based cell-pack topology is proposed, in this study, the authors have designed an integrated module by amalgamation of power electronics and battery cells. The application of which can be done in a synchronous bidirectional DC-DC converter with the help of digital control techniques. As a result, the power rating and voltage stress associated with power electronics are destroyed, and it provides the capability to have active control over thermal distribution, improving safety and reliability during the equalization procedure.

3.1.3 Direct Cell-Cell Topology

The working principle of direct cell-cell topological structure is based on a charge transferring strategy from one cell to another cell [41–44]. In this technique, the transfer of energy can take place directly between the cells with highest to the lowest SOC level by controlling a pair of switches in every working time slot. As a consequence, for b serially connected cells in a battery pack, a standard equalizer e_{SC} is used throughout the pack to independently control b sets of switches as shown in Fig. 3.7. In this case, only one pair of switches is turned on during each working time slot. The other switches, on the other hand, must be turned off to allow $epsilon_i$

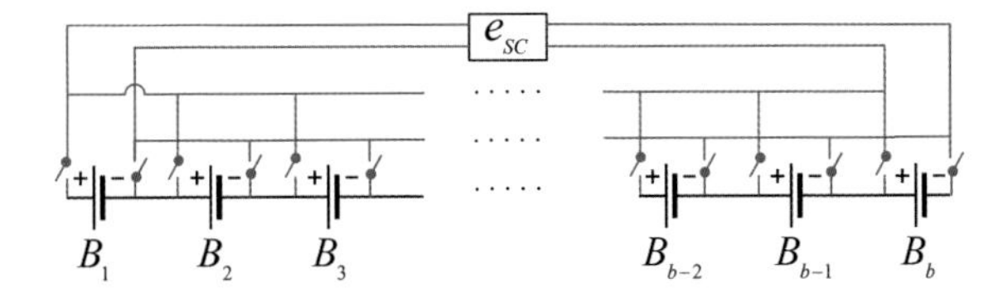

Fig. 3.7 Direct cell-cell battery charge equalization topology

energy to be transferred among any selected cells at any point in the pack using the defined charge transferring strategy.

The dynamics of cell equalizing system of a direct cell-cell topological structure for a battery pack made up of b series—connected cells could be captured by the following mathematical model:

$$x_i(k+1) = x_i(k) + d\epsilon_i(k), i = 1, ..., b \tag{3.11}$$

where $x_i(k)$ defines the SOC level of the cell B_i. It is to be noted that $\epsilon_0(k) = \epsilon_b(k) = 0$, and $\epsilon_i(k)(1 \leq i \leq b-1)$ is:

$$\epsilon_i(k) = (k_i + k_i')u_i(k)T \tag{3.12}$$

with

$$\begin{cases} k_i = 0, k_i' = -1 & \text{if } x_i(k) = \text{Max}(x(k)) \\ k_i = 1, k_i' = 0 & \text{if } x_i(k) = \text{Min}(x(k)) \end{cases}$$

in which $u_i(k) \geq 0$ is the controlled equalizing current through the equalizer e_{SC}.

Many research works published in the literature have incorporated direct cell-cell topology to enhance the cell equalizing system [10, 43, 45, 46]. In [10], the authors used a new active equalizer for transferring the energy between cells, along with a supercapacitor as an energy tank. A digitally-controlled switching converter is propounded for this reason, which would deliver additional energy from the cell with the highest SOC level to a capacitor bank and afterwards transmit it to the cell with the lowest SOC level. Consequently, the experimental findings demonstrate an magnificent improvement throughout the equalizing system efficiency of greater than 90%. In [43], a multi-winding transformer is used to propose a novel balancing control technique that utilizes the direct cell-cell equalization topology. As a matter of fact, the energy can be transmitted straightforwardly from the highest voltage rating cells to the lowest voltage rating cells by incorporating either a fly-back operation or a forward operation, resulting in a shorter equalizing path and confirming a fast equalization speed. In [45], the authors propose a different technique for balancing a series—connected battery pack based on a buck-boost converter. The propounded technique transfers charge from the highest towards the lowest charged battery cell utilizing buck-boost action, which maintains the cells at the same SOC level while employing just one magnetic element, resulting in a small size and low cost system. Based on a shuttling capacitor equalizing circuit, [7] considers a single switched capacitor cell equalizing circuit with an unique control technique to improve its equalizing performance. As a consequence of that, the equalizing system's size, cost, and time have been reduced, and also the equalizing system's efficiency improves.

3.1.4 Mixed Topology

In different equalizing layers, the mixed equalization topological structure combines and benefits from both adjacent-based and non-adjacent-based equalization topologies. During the cell equalizing procedure in the previously discussed topologies, there is inevitable duplicate charging and discharging, resulting in certain energy loss and lowering the SOH of the battery pack. The mixed equalization topology is proposed in [12, 47] to address these shortfalls while also improving cell equalizing performance results.

The mixed equalization topology is depicted in Fig. 3.8, illustrating how the battery pack with b serially connected cells is organized into m modules, M_1, M_2, ..., M_m, with each individual module containing equal n battery cells. The bottom layer was indeed based on a series-based equalization topological structure composed of various modules, in which the equalizer $e_{j,i}(1 \leq j \leq m, 1 \leq i \leq n)$ is used to transfer $\epsilon_{j,i}(k)$ quantity of energy from the battery cell B_i to $B_{i+1}(1 \leq i \leq b-1)$, or vice versa, targeting the equalizing cell having lower level of SOC. The top layer is then implemented with a module-based cell-pack equalization topology, in which the equalizer e_p is used to transfer $\epsilon_{j_p}(k)$ energy to every cell of M_j if and when the average SOC level of that individual module is less than the average SOC level of all modules. Alternatively, the e_p receives $\epsilon_{j_p}(k)$ energy from each cell in all modules. The authors in [47] proposed the cell equalizing model of the mixed topological structure for a battery pack with $b = mn$ serially connected cells, which can be represented as follows:

$$x_{j,i}(k+1) = x_{j,i}(k) + d(\epsilon_{j_p}(k) + \epsilon_{j,i-1}(k) - \epsilon_{j,i}(k)) \tag{3.13}$$

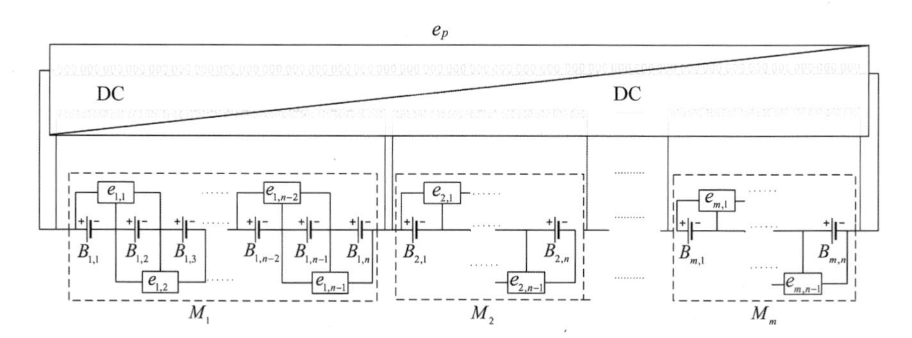

Fig. 3.8 Mixed equalization topology with series-based cell to cell equalizing in bottom layer, and module-based cell-pack in the top layer [12]

where $x_{j,i}(k)$ is the SOC level of the i-th $(1 \le i \le n)$ cell in the j-th $(1 \le j \le m)$ module; $\epsilon_{j,0}(k) = \epsilon_{j,n}(k) = 0$; $\epsilon_{j_p}(k)$ and $\epsilon_{j,i}(k)$ can be calculated as:

$$
\begin{aligned}
\epsilon_{j_p}(k) &= (k_{j_p}\tfrac{1}{b}\sum_{j=1}^{m} u_j(k) + k'_{j_p}\tfrac{u_j(k)}{n})T \\
\epsilon_{j,i}(k) &= (k_{j,i} + k'_{j,i})u_{j,i}(k)T
\end{aligned}
\tag{3.14}
$$

with

$$
\begin{cases}
k_{j_p} = 0, k'_{j_p} = 1 & \text{if } \bar{x}(k) > \bar{x}_j(k) \\
k_{j_p} = 0, k'_{j_p} = 0 & \text{if } \bar{x}(k) = \bar{x}_j(k) \\
k_{j_p} = -1, k'_{j_p} = 0 & \text{if } \bar{x}(k) < \bar{x}_j(k)
\end{cases}
$$

$$
\begin{cases}
k_{j,i} = 1, k'_{j,i} = 0 & \text{if } x_{j,i}(k) > x_{j,i+1}(k) \\
k_{j,i} = 0, k'_{j,i} = 0 & \text{if } x_{j,i}(k) = x_{j,i+1}(k) \\
k_{j,i} = 0, k'_{j,i} = -1 & \text{if } x_{j,i}(k) < x_{j,i+1}(k)
\end{cases}
$$

where $\bar{x}(k) = \frac{1}{m}\sum_{j=1}^{m} \bar{x}_j(k)$, in which $u_j(k)$, and $u_{j,i}(k)$ are the controlled equalizing current through the pack equalizer e_p, and i-th equlizer of j-th module, respectively; $\bar{x}_j(k)$ is determined in (3.4), and $\bar{x}(k)$ implies the average SOC level of the whole modules.

This topological technique is incorporated in [12], which proposes a hierarchical active equalizing framework, illustrating the application of this topological structure. The top layer is then subjected to a multi-directional multi-port converter with ideal current flow control. In consequence, the performance and advantage of their propounded equalizing topology are validated by the experimental tests.

3.2 Active Cell Equalization Topology Comparison

Throughout this section, the equalizing speed of each active cell equalization system with the aforementioned topologies is first evaluated by comparing using numerical simulations. Then an economic comparison is established between them. Following that, a discussion is presented based on the comparative results.

3.2.1 Performance Comparison

The equalizing speed is an important performance measure for the cell equalization system because it is related to many other performance indicators including loss of energy and efficiency. Multiple MATLAB/Simulink-based simulations have

Table 3.1 Simulation results for different initial cell's SOC

	$x_1(0)$ (%)	$x_2(0)$ (%)	$x_3(0)$ (%)	$x_4(0)$ (%)	$x_5(0)$ (%)	$x_6(0)$ (%)	$x_7(0)$ (%)	$x_8(0)$ (%)	$x_9(0)$ (%)	$x_{10}(0)$ (%)	$x_{11}(0)$ (%)	$x_{12}(0)$ (%)
Case 1	65	62	85	79	75	63	77	71	82	88	76	68
Case 2	80	68	52	85	77	56	82	89	72	88	69	65
Case 3	75	63	79	82	65	74	50	77	68	87	60	59
Case 4	84	73	51	78	80	86	56	54	79	64	66	52
Case 5	67	77	53	80	71	83	79	75	86	84	76	55

Table 3.2 Equalizing time for eight topologies with four different initial SOCs

Battery equalization topology	Case 2 (s)	Case 3 (s)	Case 4 (s)	Case 5 (s)
Series-based cell-cell	5239	4648	9044	5466
Module-based cell-cell	4240	4016	6255	4522
Layer-based cell-cell	3684	3013	3128	3014
Series-based cell-pack	4132	4133	3908	3683
Module-based cell-pack	3688	3125	3684	3460
Direct cell-cell	13258	12145	15278	10987
Mixed	4245	4020	4015	4531

been conducted on a battery pack consisting of 12 serially connected cells with a nominal capacity of 3.1 Ah to make a comparison of equalizing time of the eight cell equalization topologies discussed above. The controlled equalizing current has all been set as a constant of 0.5 A in [1] to better compare the equalizing time. Furthermore, all equalizers are set to be identical, but every battery pack module is assumed to comprise four cells ($n = 4, m = 3$).

The cells' initial SOCs are set at random as case 1–5 in Table 3.1. Accordingly, the simulations are performed for every topological structure of the cell equalization system, as can be seen through five illustrative simulation cases. The equalization evolutions of this battery pack using eight discussed topologies are detailed in Fig. 3.9a–g. Consequently, the layer-based topological structure has the smallest equalizing length with 2064 s, whereas the series-based and module-based topological structure have approximately the smallest equalizing time with 2901 s and 2565 s, respectively. The direct cell-cell topology, on the other hand, has the longest equalizing time with 9427 s. Furthermore, it shows that the equalizing time for the module-based cell-cell topology is 4128 s, which is 669 s shorter than the equalizing time for the series-based cell-cell topology.

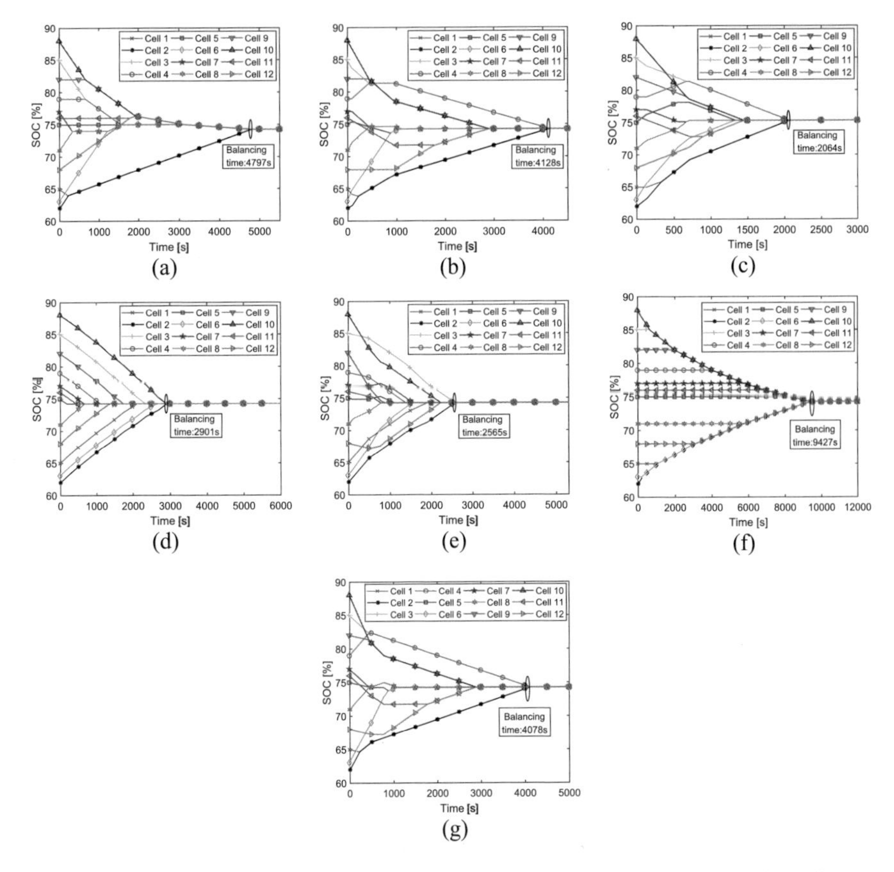

Fig. 3.9 SOC response under different cell equalizing topologies with 12 number of serially connected cells: **a** series-based cell-cell, **b** module-based cell-cell, **c** layer-based cell-cell, **d** series-based cell-pack, **e** module-based cell-pack, **f** direct cell-cell, **g** mixed

Furthermore, to investigate the equalizing speed of every topological structure, the simulations have performed with four random initial cells' SOCs to validate the performance in the first case, as can be seen in Table 3.1. In accordance, the results have shown in Table 3.2 and demonstrate that the cells' equalization could be accomplished in all cases, with the equalizing system incorporating the direct cell-cell topological structure requiring the longest equalizing duration, whereas the layer-based topological structure requires the shortest. As a consequence, the largest equalizing time in case 4 for the direct cell-cell topology is 15278 s. In case 3, that is attributed to the layer-based cell-cell topology, the smallest equalizing time is 3013 s. As a consequence, the case 1 outcomes can be confirmed further because the equalizing speed for different topologies follows the similar pattern throughout all the cases.

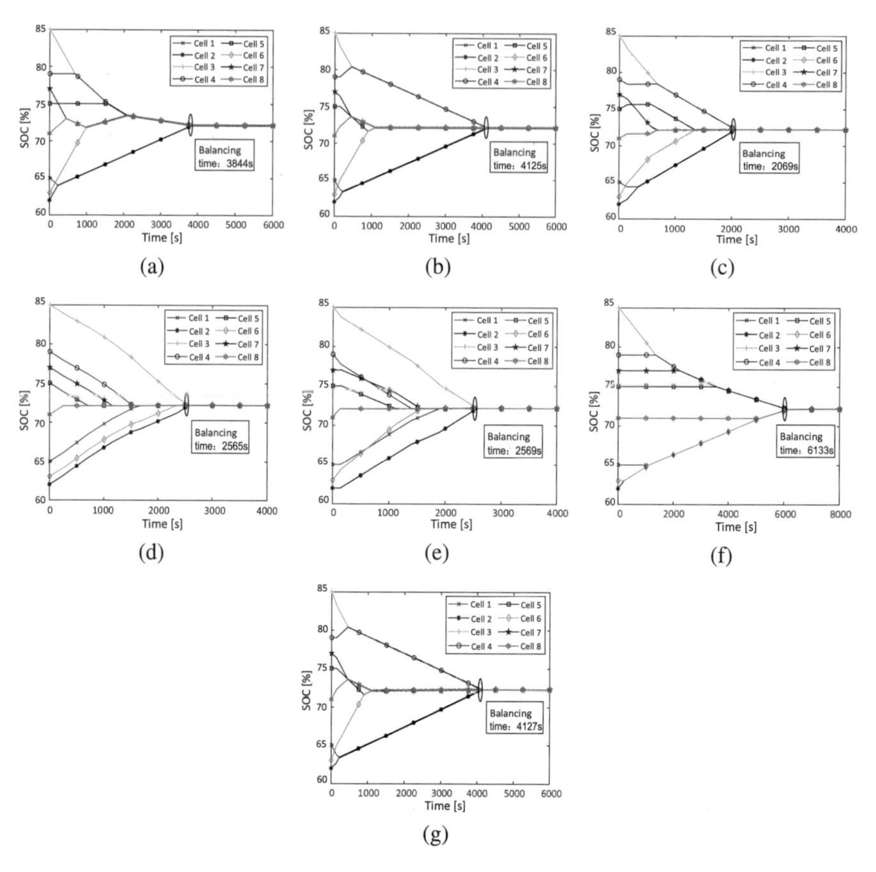

Fig. 3.10 SOC response under different cell equalizing topologies with 8 number of serially connected cells: **a** series-based cell-cell, **b** module-based cell-cell, **c** layer-based cell-cell, **d** series-based cell-pack, **e** module-based cell-pack, **f** direct cell-cell, **g** mixed

Finally, the performance of considered cell equalization topologies are investigated in Figs. 3.10 and 3.11, respectively as an illustration, which shows the cells' equalizing simulation results for the battery pack consisting of 8 and 15 cells with their initial SOCs are given as $x_i(0)(1 \leq i \leq 8) = [65\%, 62\%, 85\%, 79\%, 75\%, 63\%, 77\%, 71\%]$ and $x_i(0)(1 \leq i \leq 15) =[65\%, 62\%, 85\%, 79\%, 75\%, 63\%, 77\%, 71\%, 82\%, 88\%, 76\%, 68\%, 80\%, 73\%, 83\%]$.

In the case of an 8-cell battery pack, the findings demonstrate that the equalizing time-frame for the module-based cell-cell topology is 4125 s, which is 281 s prolonged than the series-based cell-cell topology. However, the equalizing duration of module-based topological structure is lesser than the series-based topological structure, for the battery packs having 12 and 15 cells. As a consequence, when the battery pack comprises a large number of cells, the module-based cell-cell topology performs faster than the series-based cell-cell topology. Similarly, because the bat-

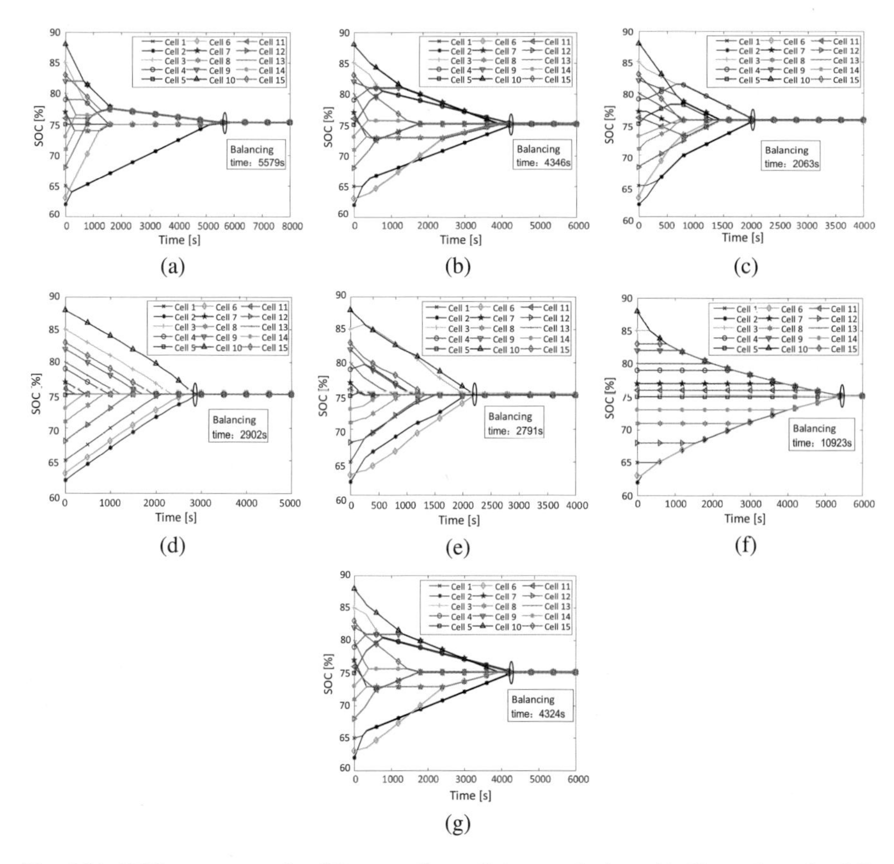

Fig. 3.11 SOC response under different cell equalizing topologies with 15 number of serially connected cells: **a** series-based cell-cell, **b** module-based cell-cell, **c** layer-based cell-cell, **d** series-based cell-pack, **e** module-based cell-pack, **f** direct cell-cell, **g** mixed

tery pack contains a large number of cells, the module-based topological structure has a shorter equalizing time than the series-based topological structure.

3.2.2 Economic Comparison

Each battery cell equalization topology listed in the literature review has substantial economic advantages as well as constraints. To summarize and compare, first determine the total number of equalizers included in each topology, which is shown as $ne_i (1 \leq i \leq 7)$ in Table 3.3. In addition, the number of electrical components associated in each equalizer is counted in a typical example of an equalizing circuit corresponding to each battery equalization topology. Consequently, for comparison, a battery pack with $b = 8$, $b = 12$, and $b = 15$ serially connected battery cells and

Table 3.3 Comparisons of active cell equalization topologies in number of equalizers and electrical components

Equalization topology	Number of equalizers in i'th topology: ne_i, $1 \leq i \leq 7$	Example of equalizing circuit	Transformer	Inductor	Switch	Capacitor
Adjacent-based	Series-based cell-cell: $ne_1 = b - 1$	Cûk converter	–	2	2	1
	Module-based cell-cell: $ne_2 = mn - 1$					
	Layer-based cell-cell: $ne_3 = \sum_{l=1}^{\lceil \log_2^b \rceil} \lceil \frac{b}{2^l} \rceil$					
Non-adjacent-based	Series-based cell-pack: $ne_4 = b$	Bidirectional multiple transformers	1	–	2	–
	Module-based cell -pack: $ne_5 = b + m$					
Direct cell-cell	Direct cell-cell: $ne_6 = 1$	Single switched capacitor	–	–	$2b$	1
Mixed	Mixed: $ne_7 = m(n - 1) + m$	Cûk converter +Bidirectional multiple transformers	1	2	4	1

Table 3.4 Comparisons of active cell equalization topologies in economic prospective

Equalization topology	Literature	Control complexity	Power loss	Size	Cost	Efficiency	Average point
Series-based cell-cell	[48–50] [3, 14, 15, 17–19] [44, 51–53]	5	1	3	5	1	3
Module-based cell-cell	[4, 6, 13, 16] [20, 21, 54, 55]	4	3	3	4	3	3.4
Layer-based cell-cell	[22–25, 27]	3	4	2.5	3	3	3.1
Series-based cell-pack	[28–31, 34, 36, 37] [35, 38, 39, 56–60]	1	4	2.5	1	3.5	2.6
Module-based cell-pack	[21, 26, 32, 40] [33, 61–68]	3	5	2	3	5	3.6
Direct cell-cell	[10, 41–43, 45] [69–72]	3	4	5	4	4	4
Mixed	[1, 12, 46, 47, 73]	3	4	1	3	4	3

Parameters of comparison: Control complexity (1: complex, 5: simple); Power loss (1: high, 5: low); Size (1: big, 5: small); Cost (1: expensive, 5: cheap); Efficiency (1: low, 5: high)

$m = 3$ modules with $n = 4$ battery cells is taken into account. Furthermore, the battery cell equalization topological structures discussed previously in this chapter are compared in Table 3.4 based on the following five factors: control complexity, power loss, size, cost, and efficiency [8, 11, 27]. To point the correspondence parameter, a numerical scale ranging from 1 to 5 is used, and finally, the average point of each topology is calculated to conclude a comparison.

From Tables 3.3 and 3.4, it can be seen that the series-based topological structure is the cheapest one and also has the simple and straightforward control structures, since the energy is transferred only between the two adjacent cells. However, it has the highest power loss and the lowest efficiency, making it unsuitable for use in a large battery pack [7, 13–15]. In turn, the module and layer-based cell-cell topologies employ a modular design structure, which provides the advantage of less power loss with greater efficiency in the case of an extended battery pack. However, control complexity and cost are higher. Even though they use the same number of equalizers, the sizes of series and module-based topological structures are nearly identical. In comparison, the size of the layer-based topology is larger than the other two [20–23].

The efficiency of the non-adjacent-based topology is higher than that of the adjacent-based topology due to its structure, which allows energy to be shuttled amongst each cell and the entire battery pack at the same time. Consequently, the battery packs have the prolonged lifetime [8, 11]. Meanwhile, the series-based cell-pack topology is the more expensive due to their high control complexity which correspondingly requires a complex control circuit, however, it has the advantages of low power loss and moderate size [28–31, 34]. However, in module-based cell-pack topological structure, the control circuit complexity, power loss, and cost are less but the size is larger. Furthermore, its suitability for larger battery packs is higher, making it suitable for high power applications [26, 32, 33].

Based on the direct cell-cell topology, a standard equalizer with switches and other electrical components such as a capacitor is used to transfer energy directly between any two cells in the battery pack. Therefore, the energy exchange between any cells at any point in the pack can be achieved, resulting in almost low cost and high efficiency cell equalization. Furthermore, this topology has the smallest size of the other topological configurations and moderate power loss [41, 42, 45]. Because undesirable charging and discharging cycles are avoided during the equalizing process, the mixed topology benefits from relatively high efficiency and low power loss [12, 47].

The following concluding remarks can be withdrawn from the performance and economic comparisons:

- The series-based cell-cell topological configuration gains the simplest control structure, however, it has the shortcomings of highest power loss and lowest efficiency among all the topological structures.
- The layer-based cell-cell topology has the fastest equalizing speed with the average economic point: 3.1 out of 5.
- The module-based cell-pack topology has perfectly fast equalizing time and nearly highest average economic point: 3.6 out of 5.
- The direct cell-cell topology gains the highest average economic point: 4 out of 5; however, the long equalizing time is the main drawback of this topological structure.

3.2.3 Discussions

It is very important to gain the benefits and avoid the drawbacks of the discussed topological configurations by selecting an appropriate option that takes into account the battery pack to be equalized. The total loss and cost of the considered equalizing system are two important factors in determining a suitable topology. Furthermore, there is always a cost-performance trade-off because higher performance is usually associated with higher costs. To choose a topology for an equalizing system, all aspects of performance, power application, and cost must be considered. The appropriate topology maximizes efficiency while minimizing power loss and cost.

The performance and economic comparisons of active cell equalizing methods provide a useful reference for equalization topology selection. As a consequence, it can be concluded that the series-based cell-cell topological configuration is appropriate for equalisation of a battery pack made up of a few serially connected cells for low power applications, as long as equalizing efficiency is not a priority over cost. The module-based cell-cell and layer-based cell-cell topological structures, on the other hand, are more advantageous for a large battery pack in high power applications, because the power loss for the series-based cell-cell is too great in this case. Furthermore, they significantly reduce the equalizing time, with the layer-based topological structure having the fastest equalizing speed. In fact, the layer-based cell-cell topology is recommended for battery packs with poor cell consistency and a high cell imbalance.

For a fast battery cell equalization while having less power loss, the series-based cell-pack, module-based cell-pack, and mixed topologies are recommended. They are also suitable for large battery packs used in high-power applications that require low power loss and a prolong battery lifespan. When a low-cost cell equalization with high efficiency is desired and equalizing speed is not critical, the direct cell-cell topological configuration is relatively advantageous for both low and high power applications.

References

1. Q. Ouyang, J. Chen, J. Zheng, H. Fang, Optimal cell-to-cell balancing topology design for serially connected lithium-ion battery packs. IEEE Trans. Sustain. Energy **9**(1), 350–360 (2018)
2. Y. Shang, Q. Zhang, N. Cui, B. Duan, C. Zhang, An optimized mesh-structured switched-capacitor equalizer for lithium-ion battery strings. IEEE Trans. Transp. Electrification **5**(1), 252–261 (2019)
3. Q. Ouyang, J. Chen, H. Liu, H. Fang, Improved cell equalizing topology for serially connected lithium-ion battery packs, in *2016 IEEE 55th Conference on Decision and Control (CDC)*, Las Vegas, USA, pp. 6715–6720 (2016)
4. W. Han, C. Zou, C. Zhou, L. Zhang, Estimation of cell soc evolution and system performance in module-based battery charge equalization systems. IEEE Trans. Smart Grid **10**(5), 4717–4728 (2019)
5. Q. Ouyang, J. Chen, C. Xu, H. Su, Cell balancing control for serially connected lithium-ion batteries, in *American control conference (ACC)*, Boston, MA, USA, 3095–3100 (2016)
6. W. Han, C. Zou, L. Zhang, Q. Ouyang, T. Wik, Near-fastest battery balancing by cell/module reconfiguration. IEEE Trans. Smart Grid **10**(6), 6954–6964 (2019)
7. H. Chen, L. Zhang, Y. Han, System-theoretic analysis of a class of battery equalization systems: mathematical modeling and performance evaluation. IEEE Trans. Veh. Technol. **64**(4), 1445–1457 (2015)

8. F. Feng, X. Hu, J. Liu, X. Lin, B. Liu, A review of equalization strategies for series battery packs: variables, objectives, and algorithms. Renew. Sustain. Energy Rev. **116**, 1–18 (2019)
9. C. Speltino, A. Stefanopoulou, G. Fiengo, Cell equalization in battery stacks through state of charge estimation polling, in *Proceedings of the 2010 American Control Conference*, Baltimore, MD, USA, pp. 5050–5055 (2010)
10. F. Baronti, G. Fantechi, R. Roncella, R. Saletti, High-efficiency digitally controlled charge equalizer for series-connected cells based on switching converter and super-capacitor. IEEE Trans. Ind. Inform. 9(2), 1139–1147 (2013)
11. J. Gallardo-Lozano, E. Romero-Cadaval, M.I. Milanes-Montero, M.A. Guerrero-Martinez, Battery equalization active methods. J. Power Sources **246**, 934–949 (2014)
12. Z. Zhang, H. Gui, D.-J. Gu, Y. Yang, X. Ren, A hierarchical active balancing architecture for lithium-ion batteries. IEEE Trans. Power Electron. **32**(4), 2757–2768 (2017)
13. N. Nguyen, S.K. Oruganti, K. Na, F. Bien, An adaptive backward control battery equalization system for serially connected lithium-ion battery packs. IEEE Trans. Veh. Technol. **63**(8), 3651–3660 (2014)
14. W. Han, L. Zhang, Y. Han, J. Li, S. Zhou, Mathematical modeling performance analysis and control of battery equalization systems: review and recent developments, in *Advances in Battery Manufacturing, Services, and Management Systems*. Wiley, pp. 281–301, 2016
15. L.R. Yu, Y.C. Hsieh, W.C. Liu, C.S. Moo, Balanced discharging for serial battery power modules with boost converters, in *2013 International Conference on System Science and Engineering (ICSSE)*, Budapest, Hungary, pp. 449–453 (2013)
16. Q. Ouyang, W. Han, C. Zou, G. Xu, Z. Wang, Cell balancing control for lithium-ion battery packs: a hierarchical optimal approach. IEEE Trans. Veh. Technol. **16**(8), 5065–5075 (2020)
17. D. Cadar, D. Petreus, T. Patarau, R. Etz, Fuzzy controlled energy converter equalizer for lithium ion battery packs, in *2011 International Conference on Power Engineering, Energy and Electrical Drives*, Torremolinos (Málaga), Spain, pp. 1–6 (2011)
18. T.T.N. Nguyen, H.-G. Yoo, S.K. Oruganti, F. Bien, Neuro-fuzzy controller for battery equalisation in serially connected lithium battery pack. IET Power Electron. **8**(3), 458–466 (2014)
19. J. Liu, Y. Chen, H.K. Fathy, Nonlinear model-predictive optimal control of an active cell-to-cell lithium-ion battery pack balancing circuit. IFAC Pap. OnLine **50**(1), 14483–14488 (2017)
20. H.-S. Park, C.-H. Kim, K.-B. Park, G.-W. Moon, J.-H. Lee, Design of a charge equalizer based on battery modularization. IEEE Trans. Veh. Technol. **58**(7), 3216–3223 (2009)
21. C.-H. Kim, M.-Y. Kim, H.-S. Park, G.-W. Moon, A modularized two-stage charge equalizer with cell selection switches for series-connected lithium-ion battery string in an HEV. IEEE Trans. Power Electron. **27**(8), 3764–3774 (2012)
22. B. Dong, Y. Han, A new architecture for battery charge equalization, in *IEEE Energy Conversion Congress and Exposition*, Phoenix, AZ, USA, vol. 2011, pp. 928–934 (2011)
23. F. Chen, J. Yuan, C. Zheng, C. Wang, Z. Li, X. Zhou, A state-of-charge based active EV battery balancing method, in *2018 2nd International Conference on Electrical Engineering and Automation (ICEEA 2018)*, Chengdu, China, pp. 70–73 (2018)
24. B. Dong, Y. Li, Y. Han, Parallel architecture for battery charge equalization. IEEE Trans. Power Electron. **30**(9), 4906–4913 (2015)
25. X. Wu, Z. Cui, X. Li, J. Du, Y. Liu, Control strategy for active hierarchical equalization circuits of series battery packs. Energies **12**(11), 2071 (2019)
26. C.-S. Lim, K.-J. Lee, N.-J. Ku, D.-S. Hyun, R.-Y. Kim, A modularized equalization method based on magnetizing energy for a series-connected lithium-ion battery string. IEEE Trans. Power Electron. **29**(4), 1791–1799 (2014)
27. M. Hoque, M. Hannan, A. Mohamed, A. Ayob, Battery charge equalization controller in electric vehicle applications: a review. Renew. Sustain. Energy Rev. **75**, 1363–1385 (2017)
28. A.M. Imtiaz, F.H. Khan, "Time shared flyback converter" based regenerative cell balancing technique for series connected li-ion battery strings. IEEE Trans. Power Electron. **28**(12), 5960–5975 (2013)
29. M. Einhorn, W. Guertlschmid, T. Blochberger, R. Kumpusch, R. Permann, F.V. Conte, C. Kral, J. Fleig, A current equalization method for serially connected battery cells using a single power converter for each cell. IEEE Trans. Veh. Technol. **60**(9), 4227–4237 (2011)

30. D. Andrea, *Battery Management Systems for Large Lithium Ion Battery Packs* (Artech house, 2010)
31. Y.-H. Hsieh, T.-J. Liang, S.-M.O. Chen, W.-Y. Horng, Y.-Y. Chung, A novel high-efficiency compact-size low-cost balancing method for series-connected battery applications. IEEE Trans. Power Electron. **28**(12), 5927–5939 (2013)
32. J. H. Lee, J. H. Lim, S. H. Park, G.-W. Moon, H. S. Park, C.-H. Kim, Two-stage charge equalization method and apparatus for series-connected battery string, US Patent 8,779,722, Jan. 2014
33. Y. Li, Y. Han, A module-integrated distributed battery energy storage and management system. IEEE Trans. Power Electron. **31**(12), 8260–8270 (2016)
34. X. Tang, C. Zou, T. Wik, K. Yao, Y. Xia, Y. Wang, D. Yang, F. Gao, Run-to-run control for active balancing of lithium iron phosphate battery packs. IEEE Trans. Power Electron. **35**(2), 1499–1512 (2020)
35. M. Hoque, M. Hannan, A. Mohamed, Voltage equalization control algorithm for monitoring and balancing of series connected lithium-ion battery. J. Renew. Sustain. Energy **8**(2), 025703 (2016)
36. S. Wang, L. Shang, Z. Li, H. Deng, J. Li, Online dynamic equalization adjustment of high-power lithium-ion battery packs based on the state of balance estimation. Appl. Energy **166**, 44–58 (2016)
37. Q. Ouyang, Z. Wang, K. Liu, G. Xu, Y. Li, Optimal charging control for lithium-ion battery packs: A distributed average tracking approach. IEEE Trans. Ind. Inform. **16**(5), 3430–3438 (2020)
38. A.M. Imtiaz, F.H. Khan, H. Kamath, A low-cost time shared cell balancing technique for future lithium-ion battery storage system featuring regenerative energy distribution, in *Twenty-Sixth Annual IEEE Applied Power Electronics Conference and Exposition (APEC)*, vol. 2011, pp. 792–799 (2011)
39. J.-W. Shin, G.-S. Seo, C.-Y. Chun, B.-H. Cho, Selective flyback balancing circuit with improved balancing speed for series connected lithium-ion batteries, in *The International Power Electronics Conference—ECCE ASIA*, Sapporo, Japan, vol. 2010, pp. 1180–1184 (2010)
40. S.-W. Lee, K.-M. Lee, Y.-G. Choi, B. Kang, Modularized design of active charge equalizer for li-ion battery pack. IEEE Trans. Ind. Electron. **65**(11), 8697–8706 (2018)
41. F. Mestrallet, L. Kerachev, J.-C. Crebier, A. Collet, Multiphase interleaved converter for lithium battery active balancing. IEEE Trans. Power Electron. **29**(6), 2874–2881 (2013)
42. Y. Shang, C. Zhang, N. Cui, J.M. Guerrero, A cell-to-cell battery equalizer with zero-current switching and zero-voltage gap based on quasi-resonant LC converter and boost converter. IEEE Trans. Power Electron. **30**(7), 3731–3747 (2015)
43. Y. Chen, X. Liu, Y. Cui, J. Zou, S. Yang, A multiwinding transformer cell-to-cell active equalization method for lithium-ion batteries with reduced number of driving circuits. IEEE Trans. Power Electron. **31**(7), 4916–4929 (2015)
44. A. F. Moghaddam, A. Van Den Bossche, An efficient equalizing method for lithium-ion batteries based on coupled inductor balancing. Electronics **8**(2), 136 (2019)
45. S.-H. Park, T.-S. Kim, J.-S. Park, G.-W. Moon, M.-J. Yoon, A new buck-boost type battery equalizer, in *Twenty-Fourth Annual IEEE Applied Power Electronics Conference and Exposition*, Washington, DC, USA, vol. 2009, pp. 1246–1250 (2009)
46. M. Daowd, M. Antoine, N. Omar, P. Van den Bossche, J. Van Mierlo, Single switched capacitor battery balancing system enhancements. Energies **6**(4), 2149–2174 (2013)
47. H.-D. Gui, Z. Zhang, D.-J. Gu, Y. Yang, Z. Lu, Y.-F. Liu, A hierarchical active balancing architecture for li-ion batteries, in *IEEE Applied Power Electronics Conference and Exposition (APEC)*, Long Beach, CA, USA, vol. 2016, pp. 1243–1248 (2016)
48. S.R. Raman, X. Xue, K.E. Cheng, Review of charge equalization schemes for li-ion battery and super-capacitor energy storage systems, in *2014 International Conference on Advances in Electronics Computers and Communications*, Bangalore, India, pp. 1–6 (2014)
49. Y.-S. Lee, M.-W. Cheng, Intelligent control battery equalization for series connected lithium-ion battery strings. IEEE Trans. Ind. Electron. **52**(5), 1297–1307 (2005)

50. Y. Shang, Q. Zhang, N. Cui, C. Zhang, A cell-to-cell equalizer based on three-resonant-state switched-capacitor converters for series-connected battery strings. Energies **10**(2), 1–15 (2017)
51. Y.-S. Lee, M.-W. Cheng, S.-C. Yang, C.-L. Hsu, Individual cell equalization for series connected lithium-ion batteries. IEICE Trans. Commun. **89**(9), 2596–2607 (2006)
52. A.F. Moghaddam, A. Van Den Bossche, An active cell equalization technique for lithium ion batteries based on inductor balancing, in *2018 9th International Conference on Mechanical and Aerospace Engineering (ICMAE)*, Budapest, Hungary, pp. 274–278 (2018)
53. A.F. Moghaddam, A. Van Den Bossche, A cell equalization method based on resonant switched capacitor balancing for lithium ion batteries, in *2018 9th International Conference on Mechanical and Aerospace Engineering (ICMAE)*, Budapest, Hungary, pp. 337–341 (2018)
54. C. Zhang, Y. Shang, Z. Li, N. Cui, An interleaved equalization architecture with self-learning fuzzy logic control for series-connected battery strings. IEEE Trans. Veh. Technol. **66**(12), 10923–10934 (2017)
55. Y. Ma, P. Duan, Y. Sun, H. Chen, Equalization of lithium-ion battery pack based on fuzzy logic control in electric vehicle. IEEE Trans. Ind. Electron. **65**(8), 6762–6771 (2018)
56. A. Khalid, A. Hernandez, A. Sundararajan, A.I. Sarwat, Simulation-based analysis of equalization algorithms on active balancing battery topologies for electric vehicles, in *Proceedings of the Future Technologies Conference*, San Francisco, USA, pp. 708–728 (2019)
57. H.-S. Park, C.-E. Kim, G.-W. Moon, J.-H. Lee, J.K. Oh, Two-stage cell balancing scheme for hybrid electric vehicle lithium-ion battery strings, in *IEEE Power Electronics Specialists Conference*, Orlando, FL, USA, vol. 2007, pp. 273–279 (2007)
58. M.A. Hannan, M.M. Hoque, S.E. Peng, M.N. Uddin, Lithium-ion battery charge equalization algorithm for electric vehicle applications. IEEE Trans. Ind. Appl. **53**(3), 2541–2549 (2017)
59. Y. Shang, B. Xia, C. Zhang, N. Cui, J. Yang, C.C. Mi, An automatic equalizer based on forward-flyback converter for series-connected battery strings. IEEE Trans. Ind. Electron. **64**(7), 5380–5391 (2017)
60. L. Liu, B. Xu, Z. Yan, W. Zhou, Y. Li, R. Mai, Z. He, A low-cost multiwinding transformer balancing topology for retired series-connected battery string. IEEE Trans. Power Electron. **36**(5), 4931–4936 (2021)
61. X. Cao, Q.-C. Zhong, Y.-C. Qiao, Z.-Q. Deng, Multilayer modular balancing strategy for individual cells in a battery pack. IEEE Trans. Energy Convers. **33**(2), 526–536 (2018)
62. M. Daowd, M. Antoine, N. Omar, P. Lataire, P. Van Den Bossche, J. Van Mierlo, Battery management system-balancing modularization based on a single switched capacitor and bi-directional dc/dc converter with the auxiliary battery. Energies **7**(5), 2897–2937 (2014)
63. F. Baronti, C. Bernardeschi, L. Cassano, A. Domenici, R. Roncella, R. Saletti, Design and safety verification of a distributed charge equalizer for modular li-ion batteries. IEEE Trans. Ind. Inform. **10**(2), 1003–1011 (2014)
64. I.-K. Baek, T.-H. Kim, C.-S. Lim, R.-Y. Kim, Modularized battery cell voltage equalization circuit using extended multi-winding transformer, in *IEEE Vehicle Power and Propulsion Conference*, Seoul, South Korea, vol. 2012, pp. 349–353 (2012)
65. H.-S. Park, C.-E. Kim, C.-H. Kim, B.-C. Kim, G.-W. Moon, J.-H. Lee, Modularized charge equalization converter with high power density and low voltage stress for HEV lithium-ion battery string, in *2007 7th Internatonal Conference on Power Electronics*, Daegu, South Korea, pp. 784–789 (2007)
66. F. Ju, W. Deng, J. Li, Performance evaluation of modularized global equalization system for lithium-ion battery packs. IEEE Trans. Autom. Sci. Eng. **13**(2), 986–996 (2016)
67. M. Gao, J. Qu, H. Lan, Q. Wu, H. Lin, Z. Dong, W. Zhang, An active and passive hybrid battery equalization strategy used in group and between groups. Electronics **9**(10), 1–21 (2020)
68. H. Nazi, E. Babaei, A modularized bidirectional charge equalizer for series-connected cell strings. IEEE Trans. Ind. Electron. **68**(8), 6739–6749 (2021)
69. A.F. Moghaddam, A. Van Den Bossche, Multi-winding equalization technique for lithium ion batteries for electrical vehicles, in *2018 7th International Conference on Renewable Energy Research and Applications (ICRERA)*, Paris, France, pp. 139–143 (2018)

70. M.-Y. Kim, C.-H. Kim, J.-H. Kim, G.-W. Moon, A chain structure of switched capacitor for improved cell balancing speed of lithium-ion batteries. IEEE Trans. Ind. Electron. **61**(8), 3989–3999 (2014)
71. S.-H. Park, K.-B. Park, H.-S. Kim, G.-W. Moon, M.-J. Youn, Single-magnetic cell-to-cell charge equalization converter with reduced number of transformer windings. IEEE Trans. Power Electron. **27**(6), 2900–2911 (2011)
72. V.-L. Pham, V.-T. Duong, W. Choi, High-efficiency active cell-to-cell balancing circuit for lithium-ion battery modules using LLC resonant converter. J. Power Electron. **20**(4), 1037–1046 (2020)
73. Y. Shang, N. Cui, B. Duan, C. Zhang, A global modular equalizer based on forward conversion for series-connected battery strings. IEEE J. Emerging Sel. Top. Power Electron. **6**(3), 1456–1469 (2018)

Chapter 4
Optimal Active Cell Equalizing Topology Design

4.1 Cell Equalizing System

Equalizing currents: The current from the i-th (j-th) cell to its linked l-th ICE, designated as $I_{il}(k)$ ($I_{jl}(k)$), is regarded as the regulated equalizing current for the scenario where energy is transmitted from the i-th (j-th) cell to the j-th (i-th) cell (Fig. 4.1). The energy transfer efficiency of the l-th ICE, satisfying $0 < \beta_l \leq 1$, determines the current from the l-th ICE to the linked j-th (i-th) cell, which is $\beta_l I_{il}(k)$ ($\beta_l I_{jl}(k)$). As a result, the unifying expression for the equalizing currents for the i-th cell and the j-th cell through the l-th ICE is as follows:

$$
\begin{aligned}
I_{eq_{il}}(k) &= k_l I_{il}(k) + k_l' \beta_l I_{jl}(k) \\
I_{eq_{jl}}(k) &= k_l \beta_l I_{il}(k) + k_l' I_{jl}(k)
\end{aligned}
\tag{4.1}
$$

with

$$
k_l = \begin{cases} 1, & \text{energy transferred from cell } i \text{ to cell } j \\ 0, & \text{otherwise} \end{cases}
$$

$$
k_l' = \begin{cases} 1, & \text{energy transferred from cell } j \text{ to cell } i \\ 0, & \text{otherwise} \end{cases}
$$

where $I_{eq_{il}}(k)$ and $I_{eq_{jl}}(k)$ are the equalizing currents for the i-th cell and the j-th cell through the l-th ICE, respectively. Energy can only flow in one direction at a time, hence $k_l k_l' = 0$ is satisfied. The current flow is referred to as positive if it flows from the cell i to the cell j and negative if it does not when considering the i-th and j-th cells with $i < j$. A positive flow in this case indicates that $k_l = 1$ and $k_l' = 0$, whereas a negative flow suggests that $k_l = 0$ and $k_l' = 1$. Therefore, it may be said that

$$
u_l(k) = \begin{cases} I_{il}(k), & u_l(k) \geq 0 \\ I_{jl}(k), & u_l(k) < 0 \end{cases}
\tag{4.2}
$$

J. Chen et al., *Equalization Control for Lithium-Ion Batteries*,
https://doi.org/10.1007/978-981-99-0220-0_4

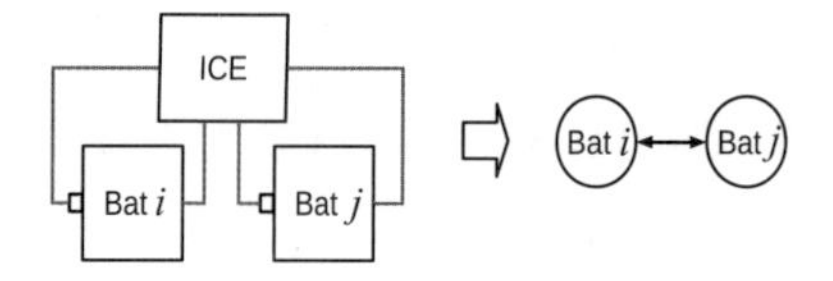

Fig. 4.1 Equivalent diagram of battery cells and their connected ICEs

where $u_l(k)$ is the controlled equalizing current of the l-th ICE. Based on (4.1)–(4.2), through their associated l-th ICE, the i-th and j-th cells' equalizing currents (for $i < j$) are

$$\begin{cases} I_{eq_{il}}(k) = u_l(k), \\ I_{eq_{jl}}(k) = \beta_l u_l(k); \end{cases} u_l(k) \geq 0 \qquad \begin{cases} I_{eq_{il}}(k) = \beta_l u_l(k), \\ I_{eq_{jl}}(k) = u_l(k), \end{cases} u_l(k) < 0. \tag{4.3}$$

4.1.1 *Equalizing System Model*

The terminal current of the i-th battery cell in an n-modular serially linked battery pack with m ICEs is calculated as follows:

$$I_{B_i}(k) = I_s(k) + \sum_{l=1}^{m} c_{il} I_{eq_{il}}(k), \quad \text{for } 1 \leq i \leq n. \tag{4.4}$$

Here, $I_{B_i}(k)$ is the terminal current of the i-th cell, which is positive (negative) depending on whether the cell is discharging or charging; $I_s(k)$ is the current flowing through the battery pack from an external power source or from a load, and because of the serial connection, it is the same for all cells; the equalizing current for the i-th cell through the l-th ICE is $I_{eq_{il}}(k)$; the coefficient, c_{il}, is given as follows:

$$c_{il} = \begin{cases} 1, & \text{cell } i \text{ connected with ICE } l \text{ and } i < j \\ -1, & \text{cell } i \text{ connected with ICE } l \text{ and } i > j \\ 0, & \text{cell } i \text{ is not connected with ICE } l. \end{cases} \tag{4.5}$$

The other cell's index was linked to the l-th ICE by j. The discrete-time dynamics of the i-th ($1 \leq i \leq n$) cell's SOC may be estimated as follows since the battery cell's SOC is a quantity determined as the proportion of the capacity that is still available to that of its nominal capacity:

$$SOC_i(k+1) = SOC_i(k) - \varphi_i I_{B_i}(k) \tag{4.6}$$

with the auxiliary variable $\varphi_i = \eta T_0/(3600 Q_i)$. In addition, $SOC_i(k)$ is the i-th cell's SOC, Q_i is the i-th cell's capacity in Ampere-hours, T_0 is the sampling period. The variable η denotes the Coulombic efficiency, which satisfies $\eta = 1$ for discharging and $\eta = \eta_c \leq 1$ for charging with η_c equal to the ratio between the cell's discharging and charging capacities [1]. Based on (4.3)–(4.6), the following may be used to depict the model of the cell equalizing system for an n-modular serially linked battery pack:

$$x_i(k+1) = x_i(k) - \varphi_i \sum_{l=1}^{m} b_{il} c_{il} u_l(k) - \varphi_i d(k) \tag{4.7}$$

where $x_i(k)(1 \leq i \leq n)$ represents the i-th battery cell's SOC, $u_l(k)(1 \leq l \leq m)$ is the controlled equalizing current of the l-th ICE, $d(k)$ represents the external current of the battery pack $I_s(k)$, and b_{il} is a coefficient that satisfies

$$b_{il} = \begin{cases} 1, & \text{cell } i \text{ connected with ICE } l \text{ and} \\ & (u_l(k) \geq 0, i < j \text{ or } u_l(k) < 0, i > j) \\ \beta_l, & \text{cell } i \text{ connected with ICE } l \text{ and} \\ & (u_l(k) < 0, i < j \text{ or } u_l(k) \geq 0, i > j) \\ 0, & \text{cell } i \text{ is not connected with ICE } l. \end{cases} \tag{4.8}$$

4.1.2 Consensus-Based Cell Equalizing Algorithm Design

Since SOC estimate has been well researched in the literature, including [2–4], it is assumed that the cells' SOCs in the serially linked battery pack are known. The controlled equalizing current of the suggested consensus-based cell equalization method may be described as follows for l-th $(1 \leq l \leq m)$ ICE linking the i-th and j-th cells:

$$u_l(k) = \alpha(k)(x_i(k) - x_j(k)) \tag{4.9}$$

where $\alpha_m \leq \alpha(k) \leq \alpha_M; \alpha_m,\ \alpha_M > 0$ is satisfied, and $\alpha(k)$ represents a positive real gain that is bounded. It should be noted that (4.9) adopts the form that is most effective and practical for execution. If appropriate, it can be generalized to proportional functions, saturation functions, or other expressions. The model of the serially linked battery pack (4.7) may be expressed as follows using the intended input in (4.9):

$$\begin{aligned} x(k+1) &= x(k) - \Psi \bar{C} u(k) - \Psi \xi(k) \\ u(k) &= \alpha(k) C^T x(k) \end{aligned} \tag{4.10}$$

where $x(k) \triangleq [x_1(k), \ldots, x_n(k)]^T \in \mathbb{R}^n$; $u(k) \triangleq [u_1(k), \ \ldots, u_m(k)]^T \in \mathbb{R}^m$; $\Psi \triangleq \mathrm{diag}\{\varphi_1, \ldots, \varphi_n\} \in \mathbb{R}^{n \times n}$ with $\mathrm{diag}\{\cdot\}$ the diagonal matrix; $C \triangleq [c_{il}] \in R^{n \times m}$ with c_{il} in (4.8) and $\bar{C} \triangleq [b_{il} c_{il}] \in R^{n \times m}$ with b_{il} in (4.8), respectively; $\xi(k) \triangleq [d(k), \ldots, d(k)]^T \in \mathbb{R}^n$. By substituting the input to the cell equalizing model, (4.10) can be represented as follows:

$$x(k+1) = A(k)x(k) - \Psi \xi(k) \tag{4.11}$$

with

$$A(k) = I_{n\times n} - \alpha(k)\Psi\bar{C}C^T.$$

In this case, $I_{n\times n}$ represents the identity matrix of dimension n. The convergence feature of the aforementioned SOC equalizing method is encapsulated in the following theorem.

Theorem 4.1 *If the following two requirements are met, the SOC differences between the cells in the n-modular serially linked battery pack can approach uniformly constrained under the protocol (4.9):*

1. *The graph G of the cell equalizing system is linked since each cell is connected to at least one ICE, and the n-modular battery pack uses at least n − 1 ICEs.*
2. $(1-\sqrt{1-\lambda_M})/\lambda_M < \alpha(k) < (1+\sqrt{1-\lambda_M})/\lambda_M$, *where* λ_M *is the largest eigenvalue of the positive definite matrix* $\bar{C}^T\Psi C$.

Proof A Lyapunov candidate designated $V(k) \in \mathbb{R}$ is chosen as the Lyapunov candidate owing to the Lyapunov function in [5] as follows:

$$V(k) = s^T(k)s(k) \tag{4.12}$$

where $s(k) = C^T x(k)$, and the graph G's incidence matrix, C, can be thought of as previously described in (4.10). When the SOCs of all the battery cells are the same, a manifold is said to exist as given below:

$$\Gamma \triangleq \{x \in R^n | x_1 = x_2 = \cdots = x_n\}. \tag{4.13}$$

Since the cell equalizing system's graph G is linked, it can be determined that $V(k) \geq 0$ and $V(k) = 0$ if and only if $x \in \Gamma$, i.e., $s = 0$ [5]. Based on (4.11) and (4.12), it yields

$$\begin{aligned}\Delta V(k) &= V(k+1) - V(k)\\ &= -s^T(k)(2\alpha(k)I_{m\times m} - \alpha^2(k)\bar{C}^T\Psi C)\Pi s(k)\\ &\quad - 2x^T(k)A^T(k)CC^T\Psi\xi(k) + \xi^T(k)\Psi CC^T\Psi\xi(k)\end{aligned} \tag{4.14}$$

with $\Pi = C^T\Psi\bar{C}$. Since $A(k) \preceq I_{n\times n}$ from (4.11) with $\preceq$ denoting the matrix inequality, using the Cauchy-Schwarz inequality [6], it is possible to determine that

$$\begin{aligned}&-2x^T(k)A^T(k)CC^T\Psi\xi(k)\\ &\quad \leq \lambda_m s^T(k)s(k) + 1/\lambda_m\xi^T(k)\Psi CC^T\Psi\xi(k)\\ &\quad \leq s^T(k)\Pi s(k) + 1/\lambda_m\xi^T(k)\Psi CC^T\Psi\xi(k)\end{aligned} \tag{4.15}$$

where $\lambda_m > 0$ is the positive definite matrix Π's minimal eigenvalue. The following may be inferred from (4.14):

$$\Delta V(k) \leq -p(k)s^T(k)\Pi s(k) + M_{11} \tag{4.16}$$

with $\quad p(k) = 2\alpha(k) - \alpha^2(k)\lambda_M - 1, \quad M_{11} = (1 + 1/\lambda_m)||C^T\Psi\xi(k)||^2.$

Since $(1-\sqrt{1-\lambda_M})/\lambda_M < \alpha(k) < (1+\sqrt{1-\lambda_M})/\lambda_M$, $p(k)\Pi$ is positive definite. From (4.16), $\Delta V(k)$ is negative outside the set $\{||s|| \le \sqrt{M_{11}/(p(k)\lambda_m)}\}$. Until the solutions enter the set $\{||s|| \le \sqrt{M_{11}/(p(k)\lambda_m)} + \delta\}$ with a tiny positive value δ, the Lyapunov function $V(k)$ will drop monotonically, and the solutions cannot exit the set beyond that point. The cells' SOCs are uniformly bounded around the manifold Γ with the bound $||s|| \le \sqrt{M_{11}/(p(k)\lambda_m)} + \delta$ according to the boundedness analysis in [7], which is equivalent to saying that the cells' SOC differences are bounded. Because M_{11} is tiny and the cells' SOC differences have a modest upper bound, the current $d(k)$ is small enough in practice compared to the capabilities of the cells in Ampere-hour Q_i. Furthermore, $C^T\Psi\xi(k) = 0_m$ and $M_{11} = 0$ for the scenario when the battery pack's cell capacities are equal, where 0_m is a zero column vector of dimension m. The SOC differences between the cells can then converge to zero. The condition $M_{11} = 0$ and the global consensus of the cell SOCs are both reached when the battery pack is in standby mode with $\xi(k) = 0_n$.

4.2 Design of the Optimal Equalizing Topology

Here, a convergence proof is used to illustrate a consensus-based cell equalizing technique and assess its efficacy. We explore expediting the equalization process by including a number of ICE-based edges in the graph to increase the crucial time efficiency. Given a set number of ICEs, the challenge of where to add them in order to maximize the reduction of the equalization time is both interesting and important because doing so will raise the cost of the entire cell equalizing system. Based on the findings, this study provides another important step toward effective cell equalizing and adds fresh insights to existing research, including [8–16].

4.2.1 Equalizing Time

The cell equalizing problem may be translated as a distributed averaging problem since the cell equalizing system is depicted as a MAS. The equalizing time may be defined as follows to assess the cell equalizing speed:

$$T(\epsilon) = \min\{\tau : \|x(k) - \bar{x}(k)\| \le \epsilon, \forall kT_0 \ge \tau\} \tag{4.17}$$

for all $x(0) \ne \bar{x}(0)$, where $||\cdot||$ denotes the 2-norm; $\bar{x}(k) = \frac{1}{n}1_n 1_n^T x(k)$ stands for the average vector that $x(k)$ converges to, with 1_n the vector of n ones; $x(0)$ is the cells' initial SOC vector and T_0 is the sampling period; ϵ is the maximum tolerant difference between the cells' SOCs and the average SOC. Since the energy transfer efficiency β_l in (4.1) is practically close to 1, assuming that $\beta_l = 1$ results in $\bar{C} = C$.

Using (4.11), it is possible to determine that

$$A(k) = I_{n\times n} - \alpha(k)\Psi CC^T. \tag{4.18}$$

The cell equalizing system's linked graph CC^T has an eigenvalue of 0, and its corresponding eigenvector is 1_n. This results in $A(k)1_n = 1_n$. By assuming that $\Psi\xi(k) = 0_n$, the influence of $\Psi\xi(k)$ on the averaging problem may be ignored. It is possible to deduce the dynamics from (4.11) and (4.18):

$$\begin{aligned} x(k) - \bar{x}(k) &= (A(k-1) - \tfrac{1}{n}1_n 1_n^T)x(k-1) \\ &= (A(k-1) - \tfrac{1}{n}1_n 1_n^T)(x(k-1) - \bar{x}(k-1)) \\ &= (A(k-1) - \tfrac{1}{n}1_n 1_n^T)\cdots(A(0) - \tfrac{1}{n}1_n 1_n^T) \times (x(0) - \bar{x}(0)). \end{aligned} \tag{4.19}$$

Because $0 < \alpha_m \le \alpha(k) \le \alpha_M$, it is deduced that $A(k) \preceq I_{n\times n} - \alpha_m\varphi_m CC^T \triangleq A_m$, where φ_m is the minimum diagonal element of Ψ. Then, it yields

$$\left\|(A(k-1) - \tfrac{1}{n}1_n 1_n^T)\cdots(A(0) - \tfrac{1}{n}1_n 1_n^T)\right\| \le \left\|A_m - \tfrac{1}{n}1_n 1_n^T\right\|^k. \tag{4.20}$$

From (4.19) and (4.20), when the difference between the cells' SOC and the average SOC $\|x(k) - \bar{x}(k)\| \le \epsilon$ after k sampling periods, it should satisfy that

$$\left\|(A_m - \tfrac{1}{n}1_n 1_n^T)^k(x(0) - \bar{x}(0))\right\| \le \left\|A_m - \tfrac{1}{n}1_n 1_n^T\right\|^k \|x(0) - \bar{x}(0)\| \le \epsilon. \tag{4.21}$$

Since $\rho(A_m - \frac{1}{n}11^T) < 1$ with $\rho(\cdot)$ the spectral radius [17], from (4.17) and (4.21), it can be deduced that the equalizing time satisfies

$$T(\epsilon) \le \frac{\log(\|x(0)-\bar{x}(0)\|)-\log(\epsilon)}{-\log(\rho(A_m - \frac{1}{n}11^T))}T_0 \propto \rho(A_m - \tfrac{1}{n}11^T). \tag{4.22}$$

where the logarithmic function is identified as $\log(\cdot)$. Using (4.18), (4.20), and utilizing [17], we can write the following:

$$\lambda_i(A_m) = 1 - \alpha_m\varphi_m\lambda_{n-i+1}(CC^T) \tag{4.23}$$

where $\lambda_1(\cdot) \le \lambda_2(\cdot) \le \cdots \le \lambda_n(\cdot)$ are the matrix's eigenvalues arranged in ascending order of magnitude. Since $\alpha_m\varphi_m\lambda_i(CC^T)$ is much less than 1, $0 \le \lambda_i(A_m) \le 1$ for $1 \le i \le n$. Referring to [17], it is obtained that

$$\rho(A_m - \tfrac{1}{n}11^T) = \lambda_{n-1}(A_m) = 1 - \alpha_m\varphi_m\lambda_2(CC^T). \tag{4.24}$$

The equalizing time $T(\epsilon)$ from (4.18), (4.22), and (4.24) can be shortened by increasing a(k) or the second lowest eigenvalue of CC^T. While a high $\alpha(k)$ can be used to shorten the equalizing time, it is still constrained by the highest equalizing current permitted by the ICE. In order to raise $\lambda_2(CC^T)$ and decrease the overall equalization time, we concentrate on the more intriguing and significant subject of how to change the topology of the equalization system. Moreover, $L = CC^T$ is known as the Laplacian matrix in graph theory. The algebraic connectivity of the graph G in [18], which is a gauge of how well-connected a graph is, is also known as its second lowest eigenvalue, $\lambda_2(L)$. Therefore, it may be inferred that a network with greater algebraic connectedness $\lambda_2(L)$ will have the cell equalizing system converge more quickly.

4.2.2 Traditional Cell Equalizing Topology

Figure 4.2a depicts the standard setup of a serially linked battery pack with n cells and $n-1$ ICEs . A undirected graph $G_{con} \triangleq (\nu, \varepsilon)$ as shown in Fig. 4.2b with the node set $\nu = \{\text{Bat } 1, \text{Bat } 2, \ldots, \text{Bat } n\}$ and the edge set $\varepsilon = \{\text{ICE } 1, \text{ICE } 2, \ldots, \text{ICE } n-1\}$ can be used to depict the topology of the system as a whole. The equalizing charge may be transported from one cell to any other cells in the battery pack through their associated ICEs since this undirected graph G_{con} is fixed and connected because there is a path between each pair of two unique nodes. The formula for its Laplacian matrix, $L(G_{con}) \in \mathbb{R}^{n \times n}$, is as follows:

$$L(G_{con}) = \begin{bmatrix} 1 & -1 & 0 & \cdots & 0 \\ -1 & 2 & -1 & \cdots & 0 \\ \cdots & \cdots & \cdots & \cdots & \cdots \\ 0 & 0 & 0 & \cdots & 1 \end{bmatrix}_{n \times n}. \tag{4.25}$$

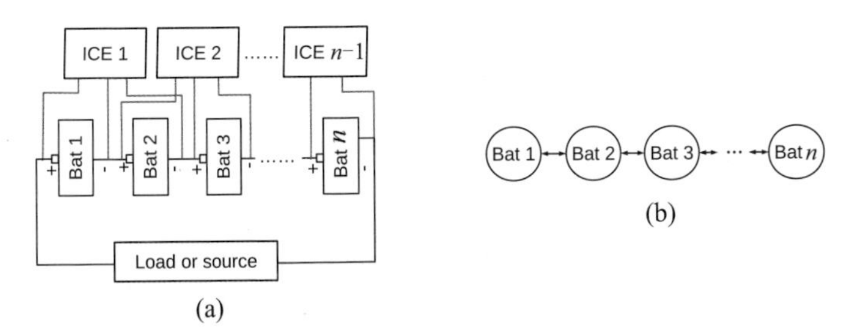

Fig. 4.2 **a** Traditional architecture of the cell equalizing system, **b** Traditional graph of the cell-equalizing system's topology

4.2.3 Position Identification of the Added ICEs for Reducing the Equalizing Time

The issue of decreasing the cell equalizing time at the maximum level by adding additional ICEs for the battery pack may be recast as the following issue based on (4.22) and (4.24). It is necessary to add γ edges ε_a from a set of m_c candidate edges ε_c to the traditional graph $G_{con} = (\nu, \varepsilon)$ shown in Fig. 4.2b that result in the highest increase in the algebraic connectivity of the new topological graph $G_{new} = (\nu, \varepsilon \cup \varepsilon_a)$. The group of potential edges ε_c is selected so that $\varepsilon \cap \varepsilon_c = \emptyset$. In other words, no more ICEs will be issued to connect two cells if they have previously been connected through an ICE. It can be recast, with reference to [18], to resolve issue as follows:

$$\begin{aligned} &\text{maximize } \lambda_2(L(G_{new})) \\ &\text{subject to } |\varepsilon_a| = \gamma, \ \ \varepsilon_a \subseteq \varepsilon_c \end{aligned} \tag{4.26}$$

where $\lambda_2(L(G_{new}))$ stands for the new topology G_{new}'s algebraic connectedness. The new topology graph $L(G_{new})$'s Laplacian matrix may be calculated from [18] as follows:

$$L(G_{new}) = L(G_{con}) + \sum_{l=1}^{m_c} \zeta s_l s_l^T \tag{4.27}$$

where m_c is the number of candidate edges, s_l is the edge vector of the candidate edge set, which is defined as $s_{l_i} = 1, s_{l_j} = -1$, and all other entries 0 for a candidate edge l linking nodes i and j with $i < j$; ζ is a 0–1 value that meets the condition $\zeta = \begin{cases} 1, \text{ edge } l \text{ is added} \\ 0, \text{ otherwise} \end{cases}$, then, the formula (4.26) can be formulated as a 0-1 programming problem as follows:

$$\begin{aligned} &\text{maximize } \lambda_2(L(G_{con}) + \sum_{l=1}^{m_c} \zeta s_l s_l^T) \\ &\text{subject to } \sum_{l=1}^{m} \zeta = \gamma, \ \ \zeta \in \{0, 1\} \end{aligned} \tag{4.28}$$

where ζ is the variable used for optimization. The position of the additional ICE that results in the largest acceleration of the convergence speed of the cell equalizing system may be found by solving the 0-1 programming in (4.28). It is more challenging to solve (4.28), since the computing cost grows exponentially as γ rises. In [18], the authors make the case that this issue may be resolved using a greedy heuristic, which adds the edges one at a time, in the goal of creating an essentially optimum solution with little computing expense. The method that has the largest value of $(v_i - v_j)^2$ where v is an eigenvector of the current Laplacian and v_i and v_j are the i- and j-th columns of v, is proposed as a technique to link the nodes i and j. Although the aforementioned heuristic algorithm is not the best option, it is still very practical and can be used to design a battery cell equalizing system. Since the cost will rise, only a small number of ICEs are actually added to the initial

Table 4.1 Change in algebraic connectivity when an ideal edge is added

Number of cells n	$\lambda_2(L(G_{con}))$	Nodes connected by the added edge (i,j)	$\lambda_2(L(G_{new}))$
5	0.382	(1, 5)	1.3820
15	0.0437	(1, 15)	0.1729
50	0.0039	(1, 50)	0.0158
100	0.00098688	(1, 100)	0.0039
200	0.00024674	(1, 200)	0.00098688

cell equalizing mechanism. Typically, $\gamma = 1$ may be used for practical application, and we'll demonstrate how adding only one ideal ICE to the simulation and trials can greatly shorten the equalizing time. Notably, the control complexity of the cell equalizing system following the addition of the ICE only slightly rises since we employ the distributed cell equalizing method (4.9).

With an ideal edge added, Table 4.1 compares the change in algebraic connectivity for battery packs with cell counts ranging from 5 to 200. It suggests that for battery packs with more cells, the algebraic connectedness is smaller, which is consistent with the idea that longer string battery packs require more equalizing time. After adding an edge that was carefully chosen, the algebraic connectivities nearly triple. The time it takes for the serially configured packs to balance can be drastically decreased by adding this edge. It should be emphasized that the edge that joins the string's initial and end cells is the one that is considered to be ideal. Given that it can be applied as a general rule and makes the new connection and wiring simple to install, this result will be highly helpful.

4.3 Simulation Results

For a battery pack with 15 cells, MATLAB-based simulations are run to verify the value of integrating the ideal ICEs with the established cell equalizing mechanism. The capacities of the cells, which range from 1.9 Ah to 2.1 Ah are chosen randomly with $[Q_1, \ldots, Q_{15}] = [1.94\text{Ah}, 2.03\text{Ah}, 2.07\text{Ah}, 1.93\text{Ah}, 2.06\text{Ah}, 1.97\text{Ah}, 1.99\text{Ah}, 2.01\text{Ah}, 2.05\text{Ah}, 2.04\text{Ah}, 1.96\text{Ah}, 1.95\text{Ah}, 1.91\text{Ah}, 2.02\text{Ah}, 1.92\text{Ah}]$. The cells' initial SOCs are given as case 1: $SOC(0) = [71\%, 72\%, 75\%, 72.5\%, 76\%, 78\%, 73\%, 74\%, 81\%, 75.5\%, 76.5\%, 70\%, 80\%, 77\%, 73.5\%]$, and the initial difference between the cells' SOCs and the average SOC is $||SOC(0) - S\bar{O}C(0)|| = 11.87\%$, where $S\bar{O}C(0) = \frac{1}{n}1_n 1_n^T SOC(0)$ denotes the initial average SOC vector. The gain $\alpha(k)$ is set to 40 in the proposed consensus-based cell equalizing method (4.9), and the sampling time is $T_0 = 2$ s. The energy transfer efficiency of the converter is set as 0.9. When the battery pack is in standby mode, Fig. 4.3a shows the SOC response for the conventional cell equalizing system under the developed

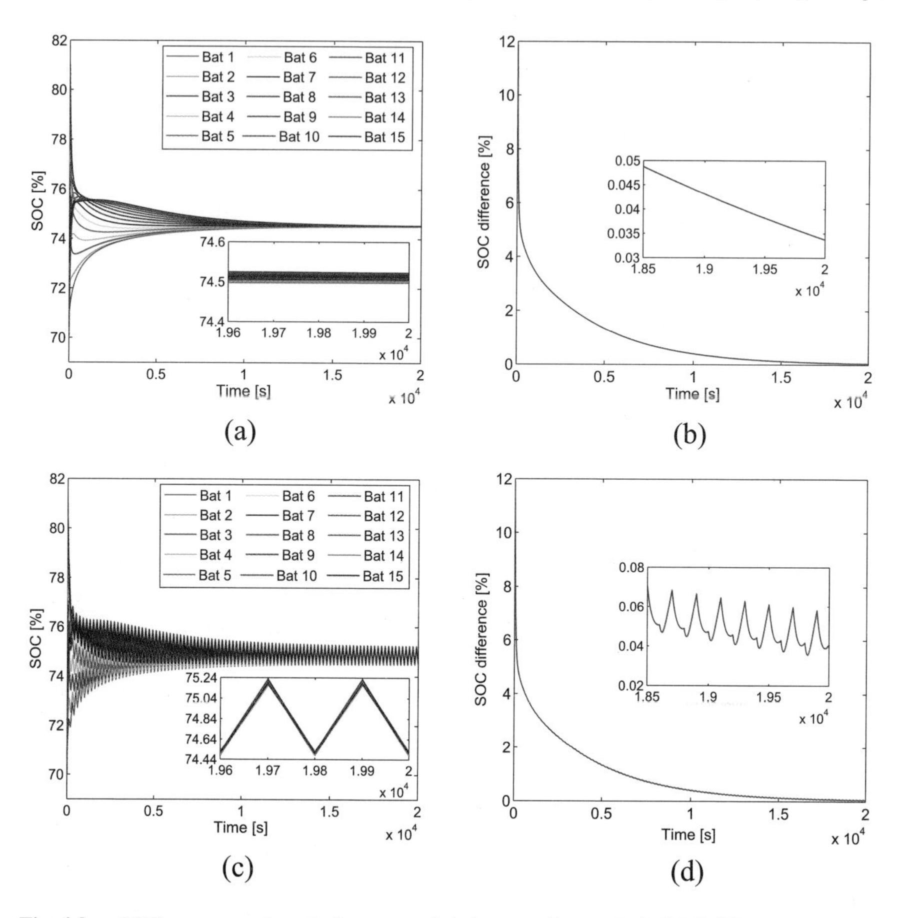

Fig. 4.3 **a** SOC response when the battery pack is in a standby mode, **b** SOC difference between the cells' SOCs and the average SOC when battery pack is in a standby mode, **c** SOC response when the battery pack is in a charging/discharging mode, **d** SOC difference between the cells' SOCs and the average SOC when the battery pack is in a charging/discharging mode

consensus-based cell equalizing algorithm, and Fig. 4.3b displays the SOC difference between the SOCs of the individual cells and the average SOC. It depicts how the SOC difference will eventually converge to zero and experience an exponential decline. The SOC response and the SOC difference for a battery pack that is alternately charged and drained at a current of 0.5 A with a period of 100 s are shown in Fig. 4.3c, d, respectively. The SOC difference for cells with various capacities will converge to a tiny bound smaller than 0.025% (oscillation magnitude), which is consistent with the theoretical study.

Now examine the scenario where one ICE is added to speed up the equalization procedure. It should be noted that the simulations may easily be expanded to include situations when numerous ICEs are added. The convergence time instant

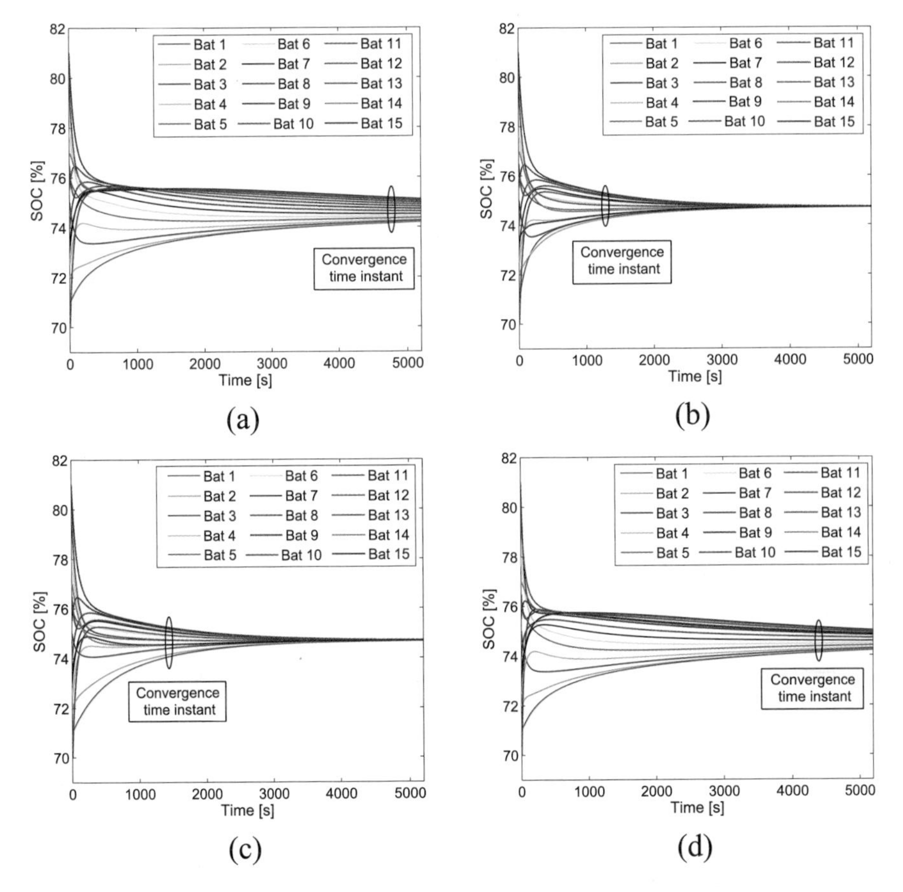

Fig. 4.4 SOC response with several topologies **a** the conventional topology, **b** one ICE linking the first and 15-th cell, **c** one ICE connecting the third and 14-th cell, **d** one ICE connecting the ninth and 12-th cell

in this context is the point in the equalizing process when the difference between the SOCs of the cells and the average SOC reaches the permissible bound of $\|SOC(k) - S\bar{O}C(k)\| \leq 1.4\%$ in (4.17). Based on the research shown above, the best ICE is chosen to link the first and fifteenth cells. The difference between the cells' SOCs under the cell equalizing control converges to the acceptable bound after 4772 s for the conventional topology of the cell equalizing system, as illustrated in Figs. 4.3a and 4.4a. As shown in Fig. 4.4b, implementing the optimum ICE reduces the equalizing time to 1286 s, which is just 26.95% of the original instance. The performance of the chosen position of the extra ICE is then shown through the use of some additional simulation data.

In Fig. 4.4c, the algebraic connectedness of the topology for the case of adding the ICE linking cells 3 and 14 is 0.1649, and the cell equalizing time is 1446 s. The cell equalizing time is 4392 s when the ICE is introduced to link cells 9 and

Table 4.2 Results of simulations for various positions of the added ICE

Cells connected by the added ICE (i, j)	$\lambda_2(L(G_{new}))$	Equalizing time
None	0.0437	4772 s in Fig. 4.4a
(1,15)	0.1729	1286 s in Fig. 4.4b
(3,14)	0.1649	1446 s in Fig. 4.4c
(9,12)	0.0544	4392 s in Fig. 4.4d

12, as illustrated in Fig. 4.4d. Given that the ninth and the twelfth cells have the greatest and lowest SOCs, respectively, at 81% and 70%, it is clear that their starting SOCs varied by the most. However, adding the ICE connecting to cells 9 and 12 does not significantly improve the speed of equalization compared to adding the best ICE. Therefore, as shown in Table 4.2, adding the ideal ICE will result in the greatest increase in cell equalizing speed. The equalizing current should be less than 10 mA/Ah [19] if a passive equalizing technology is used for cell equalization. If we stop the cell equalizing procedure when the difference between the highest SOC and the minimum SOC is less than 1.4%, a simple calculation reveals that the equalizing time for the aforementioned 15-modular battery pack is more than 34560 s. As a result, the suggested active cell equalizing system with the ideal topology requires substantially less equalizing time than the passive equalizing system.

Comparisons for different initial SOCs:
If the cell equalizing system adopts the conventional topology, the upper bound of the equalizing time for all initial SOC distributions is approximately 9244 s according to the theoretical analysis in (4.22) and (4.24). This is because the cells have a maximum capacity of 2.1 Ah. The highest bound of the equalizing time is decreased to around 2334 s by adding an ICE to connect the first and fifteenth cells, which is only nearly 25.3% of the initial time cost. The actual equalizing periods and the increases in the cell equalizing speed vary for various baseline SOC distributions. However, for all starting SOC distributions, the upper bound of the equalizing time may be decreased at a maximum level for the best cell equalizing topology. The simulations for eight additional initial cells' SOCs (case 2 to case 9) are validated in order to demonstrate that the designed optimal cell equalizing topology can improve the cell equalizing speed for any initial cells' SOCs. The initial cells' SOCs are chosen at random from the components of the previously mentioned $SOC(0)$ in Case 1 to create the initial cells' SOCs. The findings of the comparison between the equalizing timings for the conventional cell equalizing system and the cell equalizing system with the addition of a single optimum ICE linking the first and fifteenth cells are shown in Table 4.3. The longest equalizing periods, which are close to the upper bounds estimated in Table 4.3, are 8564 s in case 9 for the traditional topology and 2122 s in case 2 for the optimum topology (4.22). After adding one ICE that is optimum for the SOC distributions of all starting cells, the equalizing time may be lowered to under 2334 s. The performance of the new optimum ICEs for various starting SOC circumstances is validated by the ability to minimize the equalizing time from 6650 s in the best scenario to 974 s in the worst.

Table 4.3 Simulation comparisons of equalizing time for different initial cells' SOCs

Initial cells' SOC(%)	Traditional equalizing time(s)	Equalizing time after adding an optimal ICE(s)	Reduced equalizing time(s)
Case 1: [71, 72, 75, 72.5, 76, 78, 73, 74, 81, 75.5, 76.5, 70, 80, 77, 73.5]	4772	1286	3486
Case 2: [73, 74, 77, 81, 76, 80, 76.5, 70, 72.5, 71, 75.5, 75, 73.5, 78, 72]	3350	1474	1879
Case 3: [72, 73.5, 73, 71, 70, 72.5, 74, 76, 75, 76.5, 78, 80, 75.5, 81, 77]	7988	2112	5876
Case 4: [77, 76, 74, 73, 80, 75, 76.5, 72.5, 72, 81, 70, 71, 73.5, 75.5, 78]	2080	1106	974
Case 5: [78, 75.5, 77, 73.5, 81, 76, 76.5, 72, 71, 80, 73, 72.5, 75, 70, 74]	6014	1374	4640
Case 6: [75, 72, 81, 70, 73.5, 73, 76.5, 76, 71, 74, 75.5, 72.5, 78, 77, 80]	2914	1166	1748
Case 7: [72, 70, 73, 74, 71, 76, 75.5, 75, 81, 80, 77, 73.5, 78, 72.5, 76.5]	6634	1968	4666
Case 8: [73, 71, 72, 73.5, 70, 75, 76.5, 76, 72.5, 80, 78, 74, 77, 81, 75.5]	7412	1860	5552
Case 9: [70, 71, 72, 72.5, 73, 73.5, 74, 75, 75.5, 76, 76.5, 77, 78, 80, 81]	8564	1914	6650

4.4 Experimental Results

An experimental verification of the results previously suggested is provided in this part. The experiment uses a battery pack made up of five NCR 18650(MH12210 – 3400 mAh) lithium-ion batteries that are serially linked, as illustrated in Fig. 4.5a. The capacities of these cells are determined to be 3.1 Ah and the Coulombic efficiency η_c is 0.96 through a series of charging and discharging tests (charging/discharging the cells from the fully charged state/empty state until the cells' voltages are equal to the cut-off voltage/end-of-charge voltage). Figure 4.6a shows the connections between the cells' OCVs and SOCs. Figure 4.5b is a self-created isolated modified buck-boost converter. It is made up of two NTD6416AN-1G MOSFETs, a MAX627CPA+ dual-power MOSFET driver, a 3.46 μH WE-FB Flyback transformer, two 0.5 Ω resistors, and two MOSFETs. Figure 4.5c shows the experimental bench in detail. The LANHE battery test system charges and discharges the battery pack using the federal urban

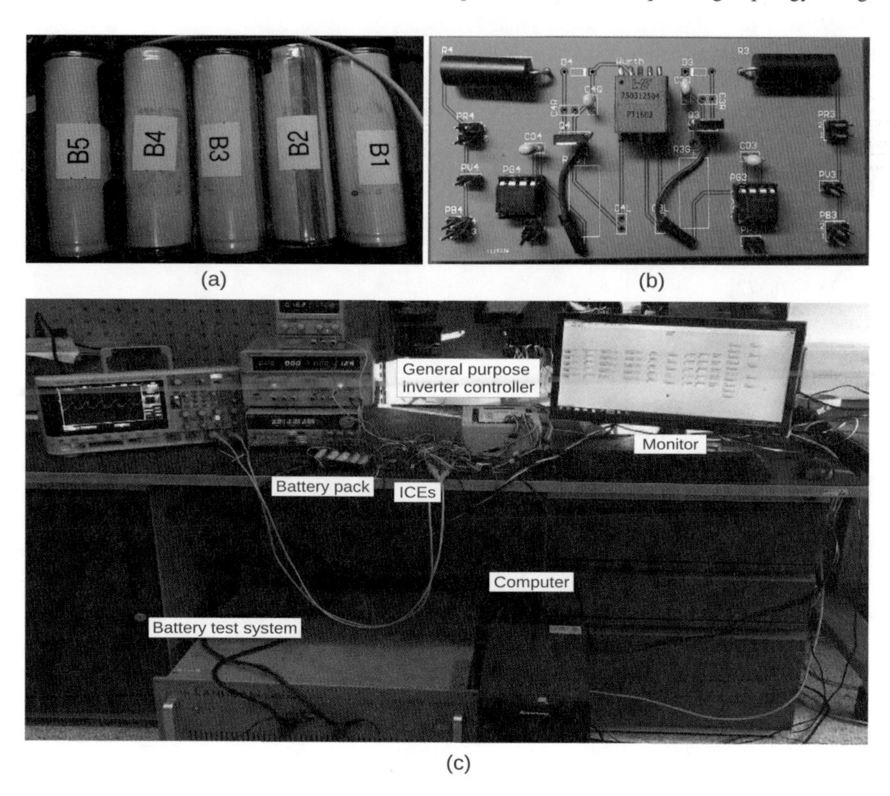

Fig. 4.5 **a** Serially connected battery pack, **b** Isolated bidirectional buck-boost converter, **c** Experimental test bench

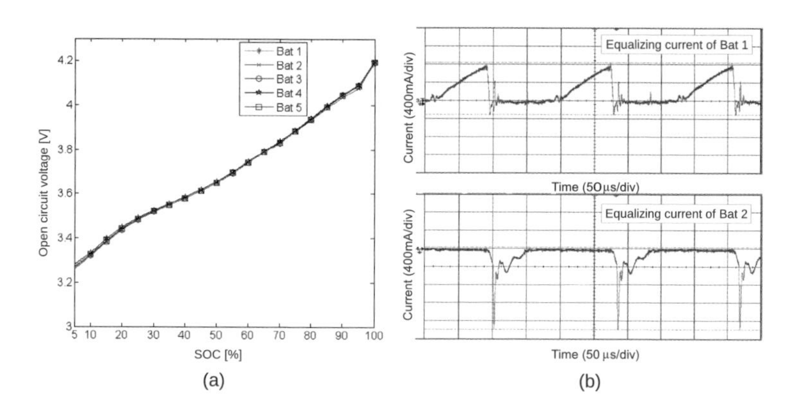

Fig. 4.6 **a** Relationship between the cells' OCVs and the SOCs, **b** Equalizing currents through the isolated bidirectional buck-boost converter

driving scheduled current curve [4], as shown in Fig. 4.7a. A National Instruments GPIC Single-Board 9683 is used to create the sampling and control signals. The regulated equalizing current is set as:

$$u_l(k) = \begin{cases} I_e, & x_i(k) - x_j(k) \geq \epsilon_1 \\ \frac{I_e}{\epsilon_1}(x_i(k) - x_j(k)), & -\epsilon_1 < x_i(k) - x_j(k) < \epsilon_1 \\ -I_e, & x_i(k) - x_j(k) \leq -\epsilon_1 \end{cases} \tag{4.29}$$

where $I_e = 0.18$ A and $\epsilon_1 = 0.5\%$ for $1 \leq l \leq m$ and $1 \leq i < j \leq n$. When the difference between the SOCs of the cells and the average SOC reaches the acceptable threshold of $c = 1.4\%$, the equalizing procedure is ended (4.17). This can successfully prevent over-equalization and lower the cost of energy for the circuits that equalize cells. Keep in mind that, depending on the situation, other acceptable values may also be provided to the regulated equalizing current.

To confirm the effectiveness of one ICE linking two cells, a test experiment is initially performed. The MOSFETs are subjected to PWM at a frequency of 124 kHz, with duty cycles of 63% and 35% for Q_1 and Q_2, respectively. The two linked cells have terminal voltages of $V_{B_1} = 3.82$ V and $V_{B_2} = 3.71$ V, respectively. The transient equalizing current curves for the cells through the linked ICE are shown in Fig. 4.6b, which was created by monitoring the voltage of the converter's resistors R_1 and R_2. The average equalizing current $I_1 = 0.1798$ A, $I_2 = 0.1084$ A, according to calculation.

In the experiment, we concentrate on the scenario in which one ICE is added, or when $\gamma = 1$. Similar steps can be taken to enable situations in which $\gamma > 1$. According to Table 4.1, the best approach to install the ICE to this five-cell battery pack is to link cells 1 and 5. The algebraic connectivity goes from 0.382 to 1.382 by adding this ICE. Another experiment is run as a comparative to examine the effects of connecting cells 2 and 4 with the ICE in a less-than-optimal way. The algebraic interconnectivity will only rise to 0.6972 in the latter scenario. After being discharged from their completely charged states, the battery cells' initial SOCs are set as case 1: SOC(1)(0) = $SOC^{(1)}(0)$ =[81%, 82%, 76%, 78%, 74%]. The equalization when the topology is conventional is shown in Fig. 4.7b. It can be observed that the SOC variations between the cells eventually converge, but it takes them around 3368 s to get there, which is too long. The equalizing time is decreased significantly to 2034 s by adding the ICE linking cells 1 and 5, as illustrated in Fig. 4.7c, which is a reduction of 1334 s when compared to the standard architecture. According to Fig. 4.7d, the cell equalizing time when the ICE is introduced to link cells 2 and 4 is about 2336 s. While this improves things somewhat, it falls short of the ideal addition technique.

It proves, in agreement with the theoretical study, that the equalisation speed grows with the algebraic connectedness of the topology. It also demonstrates how crucial it is to know how to insert the additional ICE. When the cells 1 and 5 are connected using the best addition, the convergence rate is much increased. In contrast, when cells 2 and 4 are joined, there is little improvement shown.

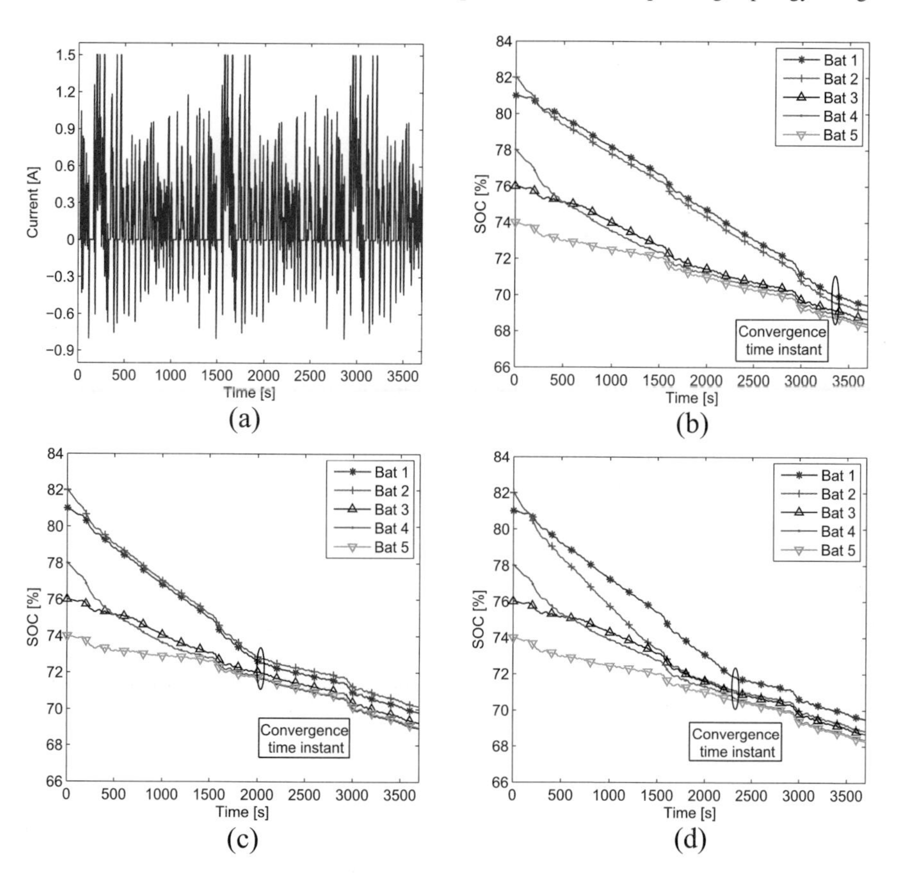

Fig. 4.7 **a** External current, **b** SOC reaction with the conventional topology, **c** SOC response with one ICE added to link cells 1 and 5, **d** SOC response with one ICE added to link cells 2 and 4

Comparison for different initial SOCs: Additional experiments are conducted whereby cells begin to balance from various initial SOCs in order to study the impact of initial cells' SOCs on the equalizing time. With the outcomes shown in Fig. 4.8 and Table 4.4, three further scenarios are tested. The cells in these cases start with case 2: $SOC^{(2)}(0) = [74\%, 81\%, 78\%, 82\%, 76\%]$, case 3: $SOC^{(3)}(0) = [81\%, 74\%, 82\%, 78\%, 76\%]$, and case 4: $SOC^{(4)}(0) = [82\%, 76\%, 78\%, 81\%, 74\%]$. It has been shown that, in every instance, adding one edge to join the first and fifth cells will drastically shorten the equalization period.

Table 4.5 compares the average SOC values at the conclusion of the cell equalizing procedure for the conventional cell equalizing system and the cell equalizing system with the addition of one ideal ICE. Since the cell equalizing system adds one ideal ICE, it requires less equalizing time and results in reduced energy transfer loss during the testing, proving that there is more energy left (higher average SOC) in the battery pack.

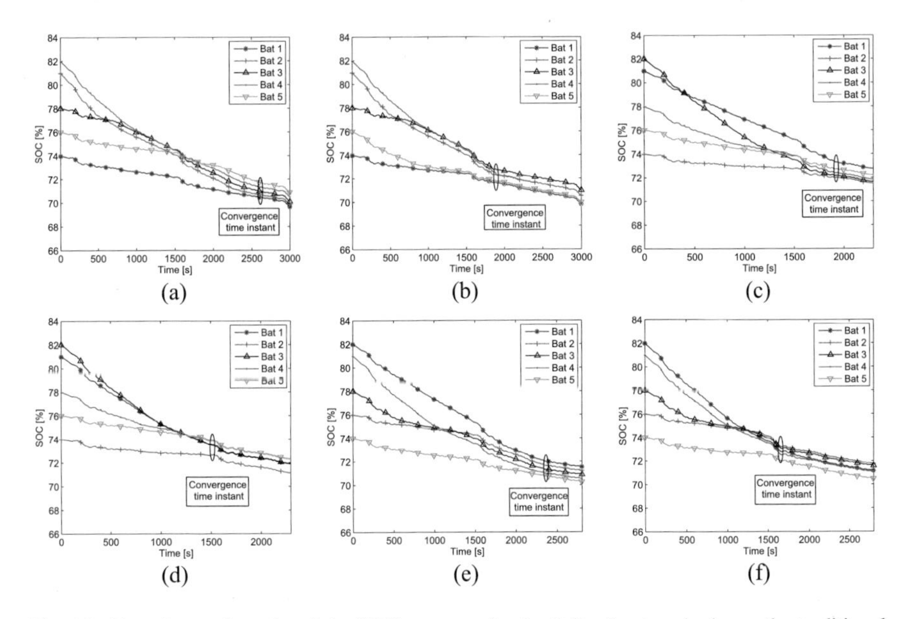

Fig. 4.8 Experimental results of the SOC response for the following topologies: **a** the traditional topology with initial SOCs of $SOC^{(2)}(0)$, **b** adding one optimal ICE with $SOC^{(2)}(0)$, **c** the traditional topology with $SOC^{(3)}(0)$, **d** adding one optimal ICE with $SOC^{(3)}(0)$, **e** the traditional topology with $SOC^{(4)}(0)$, **f** adding one optimal ICE with $SOC^{(4)}(0)$

Table 4.4 Comparisons of equalizing times for different initial SOCs

Initial cells' SOCs	Traditional (s)	New (s)	Reduced (s)
[81%, 82%, 76%, 78%, 74%]	3368	2034	1334
[74%, 81%, 78%, 82%, 76%]	2616	1884	732
[81%, 74%, 82%, 78%, 76%]	1926	1508	418
[82%, 76%, 78%, 81%, 74%]	2388	1648	740

Table 4.5 Comparisons of average SOCs after cell equalizing

	Traditional (%)	New (%)
Average SOC at 3700 s with $SOC^{(1)}(0)$	75.34	75.98
Average SOC at 3000 s with $SOC^{(2)}(0)$	75.64	75.80
Average SOC at 2300 s with $SOC^{(3)}(0)$	76.08	76.16
Average SOC at 2800 s with $SOC^{(4)}(0)$	75.66	75.96

References

1. G.L. Plett, Extended kalman filtering for battery management systems of LiPB-based HEV battery packs: part 2. Modeling and identification. J. Power Sources **134**(2), 262–276 (2004)
2. X. Lin, A.G. Stefanopoulou, Y. Li, R.D. Anderson, State of charge imbalance estimation for battery strings under reduced voltage sensing. IEEE Trans. Control Syst. Technol. **23**(3), 1052–1062 (2015)
3. H. Fang, X. Zhao, Y. Wang, Z. Sahinoglu, T. Wada, S. Hara, R.A. de Callafon, Improved adaptive state-of-charge estimation for batteries using a multi-model approach. J. Power Sources **254**, 258–267 (2014)
4. J. Chen, Q. Ouyang, C. Xu, H. Su, Neural network-based state of charge observer design for lithium-ion batteries. IEEE Trans. Control. Syst. Technol. **26**(1), 313–320 (2017)
5. T. Yang, Z. Meng, D.V. Dimarogonas, K.H. Johansson, Global consensus for discrete-time multi-agent systems with input saturation constraints. Automatica **50**(2), 499–506 (2014)
6. M. Masjed-Jamei, A functional generalization of the cauchy-schwarz inequality and some subclasses. Appl. Math. Lett. **22**(9), 1335–1339 (2009)
7. H.K. Khalil, *Nonlinear Systems* (Prentice Hall, Englewood Cliffs, NJ, USA, 2002)
8. P.A. Cassani, S.S. Williamson, Design, testing, and validation of a simplified control scheme for a novel plug-in hybrid electric vehicle battery cell equalizer. IEEE Trans. Ind. Electron. **57**(12), 3956–3962 (2010)
9. Y.-S. Lee, M.-W. Cheng, Intelligent control battery equalization for series connected lithium-ion battery strings. IEEE Trans. Ind. Electron. **52**(5), 1297–1307 (2005)
10. F. Altaf, B. Egardt, L. J. Mårdh, Load management of modular battery using model predictive control: Thermal and state-of-charge balancing. IEEE Trans. Control. Syst. Technol. **25**(1), 47–62 (2017)
11. X. Lu, K. Sun, J.M. Guerrero, J.C. Vasquez, L. Huang, State-of-charge balance using adaptive droop control for distributed energy storage systems in DC microgrid applications. IEEE Trans. Ind. Electron. **61**(6), 2804–2815 (2014)
12. L.Y. Wang, C. Wang, G. Yin, F. Lin, M.P. Polis, C. Zhang, J. Jiang, Balanced control strategies for interconnected heterogeneous battery systems. IEEE Trans. Sustain. Energy **7**(1), 189–199 (2016)
13. N. Nguyen, S.K. Oruganti, K. Na, F. Bien, An adaptive backward control battery equalization system for serially connected lithium-ion battery packs. IEEE Trans. Veh. Technol. **63**(8), 3651–3660 (2014)
14. C.-S. Lim, K.-J. Lee, N.-J. Ku, D.-S. Hyun, R.-Y. Kim, A modularized equalization method based on magnetizing energy for a series-connected lithium-ion battery string. IEEE Trans. Power Electron. **29**(4), 1791–1799 (2014)
15. P.A. Cassani, S.S. Williamson, Feasibility analysis of a novel cell equalizer topology for plug-in hybrid electric vehicle energy-storage systems. IEEE Trans. Veh. Technol. **58**(8), 3938–3946 (2009)
16. S. Yarlagadda, T.T. Hartley, I. Husain, A battery management system using an active charge equalization technique based on a DC/DC converter topology. IEEE Trans. Ind. Appl. **49**(6), 2720–2729 (2013)
17. L. Xiao, S. Boyd, Fast linear iterations for distributed averaging. Syst. Control. Lett. **53**(1), 65–78 (2004)
18. A. Ghosh, S. Boyd, Growing well-connected graphs, in *Proceedings of the 45th IEEE Conference on Decision and Control*, San Diego, CA, USA, pp. 6605–6611 (2006)
19. M. Daowd, N. Omar, P. Van Den Bossche, J. Van Mierlo, Passive and active battery balancing comparison based on matlab simulation, in *IEEE Vehicle Power and Propulsion Conference*, Chicago, IL, USA vol. 2011, pp. 1–7 (2011)

Chapter 5
Neural Network-Based SOC Observer Design for Batteries

5.1 Battery Model

An intuitive and comprehensive battery equivalent circuit model in [1] is used to accurately estimate the battery's SOC. The impacts of the cycle number and battery current on the equivalent circuit model are ignored within certain error tolerances. In general, the temperature of battery is considered to be practically constant because of the external thermal controller [2]. Equivalent circuit model of the battery having two interconnected subcircuits is depicted in Fig. 5.1, the subcircuits affect each other via controlled voltage and current sources.

The left part of the subcircuit in Fig. 5.1 is used to simulate the SOC and remaining runtime of the battery; C_b represents the stored charge in the battery when it is fully charged, and R_{sd} defines the self-discharge resistor. Furthermore, in Fig. 5.1, $V_{SOC}(t)$ indicates the battery SOCs quantitatively, and $V_{SOC}(t) \in [\,0\text{ V}, 1\text{ V}\,]$ corresponds to $0-100\%$ for the SOC. The controlled voltage source demonstrates the nonlinear mapping from the battery's SOC to the open circuit voltage $V_{OC}(t)$ as [3]:

$$V_{OC} = f(V_{SOC}) \tag{5.1}$$

where $f(\cdot)$ is a nonlinear function with the assumption that its first order derivative exist. The right part of the subcircuit in Fig. 5.1 represents the transient response and V-I curve of the battery [4]. The resistor R_0 is utilized to characterize the battery's energy losses during charge and discharge process. The RC networks (R_s, C_s) and (R_f, C_f) are used to present the battery's short-term and long-term transient responses, respectively, and R_f and C_f are larger than R_s and C_s. The resistances and capacitances in the circuit are functions of the battery's SOC. The battery's terminal voltage is denoted by $V_B(t)$. The battery current is symbolized by $I_B(t)$, which can be positive or negative depending on whether the battery is discharging or charging.

In practice, C_b can be thought of as the battery's nominal capacity, and R_{sd} can be thought of as a large constant resistor with the temperature of the battery

J. Chen et al., *Equalization Control for Lithium-Ion Batteries*,
https://doi.org/10.1007/978-981-99-0220-0_5

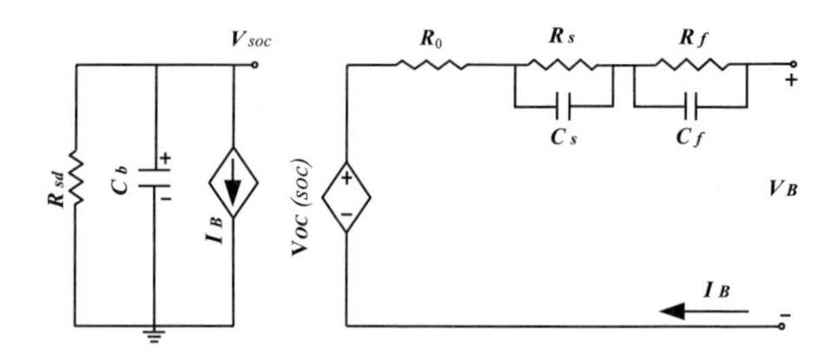

Fig. 5.1 Equivalent circuit model of the battery

varying within a small range to simplify the battery model [1]. Because the effects of capacitance change rates C_f and C_s on the battery's V-I characteristics can be ignored [5], the dynamics of the voltages across the capacitors C_b, C_s, and C_f denoted as $V_{SOC}(t)$, $V_s(t)$, and $V_f(t)$ are expressed as:

$$\begin{aligned}\dot{V}_{SOC} &= -\frac{1}{R_{sd}C_b}V_{SOC} - \frac{1}{C_b}I_B \\ \dot{V}_s &= -\frac{1}{R_s(SOC)C_s(SOC)}V_s + \frac{1}{C_s(SOC)}I_B \\ \dot{V}_f &= -\frac{1}{R_f(SOC)C_f(SOC)}V_f + \frac{1}{C_f(SOC)}I_B\end{aligned} \tag{5.2}$$

where $SOC(t)$ denotes the battery's SOC. The terminal voltage $V_B(t)$ can be formulated as

$$V_B = V_{OC} - R_0(SOC)I_B - V_f - V_s. \tag{5.3}$$

Based on (5.1), (5.2) and (5.3), the derivative of the terminal voltage is determined as:

$$\begin{aligned}\dot{V}_B = &\frac{V_f}{R_f(SOC)C_f(SOC)} - \frac{V_f}{R_s(SOC)C_s(SOC)} \\ &- \frac{V_B}{R_s(SOC)C_s(SOC)} - [\frac{R_0(SOC)}{C_s(SOC)R_s(SOC)} \\ &+ \frac{1}{C_f(SOC)} + \frac{1}{C_s(SOC)} + \frac{d(V_{SOC})}{C_b}]I_B \\ &+ \frac{f(V_{SOC})}{R_s(SOC)C_s(SOC)} - \frac{d(V_{SOC})}{R_{sd}C_b}V_{SOC}\end{aligned} \tag{5.4}$$

by assuming the derivate of the battery's current $I_B(t)$ to be negligible [6]. Furthermore, $d(V_{SOC})$ indicates the derivative of $f(V_{SOC})$ with respect to $V_{SOC}(t)$. Since the value of $V_{SOC}(t)$ ($V_{SOC}(t) \in [0, 1]$) equals to the battery's SOC ($SOC(t) \in [0, 100\%]$), $SOC(t)$ and $V_{SOC}(t)$ are interchangeable in the battery model. The battery can be modeled in the presence of modeling errors and unknown disturbances based on (5.2) and (5.4) as follows:

$$\begin{aligned}\dot{x} &= A(x)x + g(x, I_B) + \varphi(x, I_B) \\ y &= Cx\end{aligned} \tag{5.5}$$

where $x(t) \triangleq [SOC, V_f, V_s, V_B]^T \in \mathbb{R}^4$ is the system state vector; $y(t) \triangleq V_B \in \mathbb{R}$ is the output; $I_B \in \mathbb{R}$ is the measurable disturbance; $\varphi(x, I_B) \in \mathbb{R}^4$ defines the modeling errors and unmeasurable disturbances; $A(x) \in \mathbb{R}^{4\times4}$, $g(x, I_B) \in \mathbb{R}^4$, $\varphi(x, I_B)$and $C \in \mathbb{R}^{1\times4}$ are defined as:

$$A(x) = \begin{bmatrix} -a & 0 & 0 & 0 \\ 0 & -h_1(x_1) & 0 & 0 \\ 0 & 0 & -h_2(x_1) & 0 \\ 0 & h_3(x_1) & 0 & -h_2(x_1) \end{bmatrix} \tag{5.6}$$

$$g(x, I_B) = \begin{bmatrix} -bI_B \\ g_1(x_1)I_B \\ g_2(x_1)I_B \\ g_3(x_1)I_B + g_4(x_1) \end{bmatrix} \tag{5.7}$$

$$\varphi(x, I_B) = \left[0\ 0\ 0\ \psi(x, I_B)\right]^T \tag{5.8}$$

$$C = \left[0\ 0\ 0\ 1\right] \tag{5.9}$$

with $a = \frac{1}{R_{sd}C_b}$, $b = \frac{1}{C_b}$, $h_1(x_1) = \frac{1}{R_f(SOC)C_f(SOC)}$, $h_2(x_1) = \frac{1}{R_s(SOC)C_s(SOC)}$, $h_3(x_1) = \frac{1}{R_f(SOC)C_f(SOC)} - \frac{1}{R_s(SOC)C_s(SOC)}$, $g_1(x_1) = \frac{1}{C_f(SOC)}$, $g_2(x_1) = \frac{1}{C_s(SOC)}$
$g_3(x_1) = -\frac{R_0(SOC)}{C_s(SOC)R_s(SOC)} - \frac{1}{C_f(SOC)} - \frac{1}{C_s(SOC)} - \frac{d(SOC)}{C_b}$
$g_4(x_1) = \frac{f(SOC)}{R_s(SOC)C_s(SOC)} - \frac{d(SOC)}{R_{sd}C_b}SOC.$

Remark 5.1 The electrical components such as capacitances and resistances in the equivalent circuit model are assumed to be positive and bounded. Furthermore, the charging/discharging currents in the battery are bounded as BMS is protected. The states of battery system are constrained by the battery's non-extreme current. Therefore, $h_1(\cdot)$ and $h_2(\cdot)$ in (5.6) are proved to be positive and bounded. Also, $I_B(t)$, $x(t)$, $h_3(\cdot)$, and $g_i(\cdot)$ $(1 \le i \le 4)$ in (5.5)–(5.9) are all bounded.

5.2 RBF Neural Network Observer

5.2.1 Neural Network Based Nonlinear Observer Design

For estimating the SOC of batteries, a nonlinear observer incorporating an RBF neural network concept for on-line uncertainty approximation in the battery model can be designed as:

$$\dot{\hat{x}} = A(\hat{x})\hat{x} + g(\hat{x}, I_B) + \hat{\varphi}(\hat{x}, I_B) + L(y - C\hat{x}) \tag{5.10}$$

where $\hat{x}(t) \in \mathbb{R}^4$ is the estimated value of the state $x(t)$, $L \triangleq [l_1, l_2, l_3, l_4]^T$ is the gain vector of the designed observer, $\hat{\varphi}(\cdot)$ is the estimation of $\varphi(\cdot)$ in (5.5). $\hat{\psi}(\cdot)$ is an RBF neural network used to estimate the modeling errors and unmeasurable disturbance as:

$$\hat{\psi} = W^T S(\hat{z}) \tag{5.11}$$

where $\hat{z} = [\hat{x}^T, I_B]^T$ is the input vector; $W = [w_1, w_2, \ldots, w_n]^T$ represents the weight vector having n number of node in the neural network; $S(\hat{z}) = [s_1(\hat{z}), \ldots, s_n(\hat{z})]^T$ represents the activation function vector with $s_i(\hat{z})(1 \leq i \leq n)$ satisfying

$$s_i(\hat{z}) = \exp[\frac{-(\hat{z}-\mu_i)^T(\hat{z}-\mu_i)}{\eta_i^2}] \tag{5.12}$$

where $\mu_i \in \mathbb{R}^5$ is the center of the receptive field and $\eta_i \in \mathbb{R}$ is the width of a Gaussian function. The weight vector's adaptive law can be taken as [7, 8]:

$$\dot{W} = \Gamma[S(\hat{z})(y - C\hat{x}) - K_w W] \tag{5.13}$$

where $\Gamma = \Gamma^T \in \mathbb{R}^{n\times n}$ and $K_w \in \mathbb{R}^{n\times n}$ are the constant positive definite matrices to be designed.

Remark 5.2 The RBF neural network having sufficient number of nodes (neural) is shown in [9] to be approximated by a continuous function.The RBF neural network's approximation property allows the uncertain term $\psi(\cdot)$ to be replaced by

$$\psi(x, I_B) = W^{*T} S(z) + \xi \tag{5.14}$$

where W^* is the constant ideal weight vector; $z = [x^T, I_B]^T$ is the input vector; ξ represents the approximation error with the assumption $|\xi| \leq \xi_N$. Since $\psi(\cdot)$ is bounded, the ideal constant weight W^* is also bounded as:

$$\|W^*\| \leq W_M \tag{5.15}$$

where W_M is a positive constant.

Then, the nonlinear observer incorporating the RBF neural network concept in (5.10) can be modified as follows:

$$\begin{aligned}
\dot{\hat{x}}_1 &= -a\hat{x}_1 - bI_B + l_1(x_4 - \hat{x}_4) \\
\dot{\hat{x}}_2 &= -h_1(\hat{x}_1)\hat{x}_2 + g_1(\hat{x}_1)I_B + l_2(x_4 - \hat{x}_4) \\
\dot{\hat{x}}_3 &= -h_2(\hat{x}_1)\hat{x}_3 + g_2(\hat{x}_1)I_B + l_3(x_4 - \hat{x}_4) \\
\dot{\hat{x}}_4 &= h_3(\hat{x}_1)\hat{x}_2 - h_2(\hat{x}_1)\hat{x}_4 + g_3(\hat{x}_1)I_B + g_4(\hat{x}_1) \\
&\quad + W^T S(\hat{x}, I_B) + l_4(x_4 - \hat{x}_4) \\
\dot{W} &= \Gamma[S(\hat{x}, I_B)(y - C\hat{x}) - K_w W]
\end{aligned} \tag{5.16}$$

where $h_i(\cdot)$ $(1 \le i \le 3)$ and $g_i(\cdot)$ $(1 \le i \le 4)$ are all bounded. As $h_1(\cdot)$ and $h_2(\cdot)$ are positive, it can be defined that $-|h_1(\cdot)| \le -d_1$, $-|h_2(\cdot)| \le -d_2$ and $|h_3(\cdot)| \le d_3$ for positive constants d_i $(1 \le i \le 3)$. Based on (5.5), (5.14) and (5.16), the estimation error $\tilde{x}(t) \triangleq x(t) - \hat{x}(t)$ can be obtained as follows:

$$\begin{aligned}
\dot{\tilde{x}}_1 &= -a\tilde{x}_1 - l_1\tilde{x}_4 \\
\dot{\tilde{x}}_2 &= -h_1(x_1)x_2 + h_1(\hat{x}_1)\hat{x}_2 + (g_1(x_1) - g_1(\hat{x}_1))I_B - l_2\tilde{x}_4 \\
\dot{\tilde{x}}_3 &= -h_2(x_1)x_3 + h_2(\hat{x}_1)\hat{x}_3 + (g_2(x_1) - g_2(\hat{x}_1))I_B - l_3\tilde{x}_4 \\
\dot{\tilde{x}}_4 &= h_3(x_1)x_2 - h_3(\hat{x}_1)\hat{x}_2 - h_2(x_1)x_4 + h_2(\hat{x}_1)\hat{x}_4 \\
&\quad + (g_3(x_1) - g_3(\hat{x}_1))I_B + (g_4(x_1) - g_4(\hat{x}_1)) \\
&\quad + W^{*T}S(x, I_B) + \xi - W^T S(\hat{x}, I_B) - l_4\tilde{x}_4.
\end{aligned} \tag{5.17}$$

Since $h_i(\cdot)$ $(1 \le i \le 3)$, $g_i(\cdot)$ $(1 \le i \le 4)$, $x_i(t)$ $(1 \le i \le 4)$ and $I_B(t)$ are all bounded, the following expression can be deducted:

$$\begin{aligned}
&|(h_1(\hat{x}_1) - h_1(x_1))x_2 + (g_1(x_1) - \text{[illegible]}(\hat{x}_1))I_B| \le M_1 \\
&|(h_2(\hat{x}_1) - h_2(x_2))x_3 + (g_2(x_1)\,\text{[illegible]}_2(\hat{x}_1))I_B| \le M_2 \\
&|(h_3(x_1) - h_3(\hat{x}_1))x_2 + (h_2(\text{[illegible]}\, h_2(x_1))x_4| \\
&\quad + |(g_3(x_1) - g_3(\hat{x}_1))I_B + \text{[illegible]}_1) - g_4(\hat{x}_1)| \le M_3
\end{aligned} \tag{5.18}$$

where $M_i (1 \le i \le 3)$ is bounded posi[illegible]onstant. By adding and subtracting $W^{*T}S(\hat{x}, I_B)$ simultaneously, we can [illegible] the following:

$$W^{*T}S(x, I_B) + \xi - \text{[illegible]}(\hat{x}, I_B) = -\tilde{W}^T S(\hat{x}, I_B) + \varepsilon \tag{5.19}$$

where

$$\begin{aligned}
\tilde{W} &= \text{[illegible]}V^* \\
\varepsilon &\,\text{[illegible]}(S(x, I_B) - S(\hat{x}, I_B)) + \xi.
\end{aligned} \tag{5.20}$$

In (5.20), ε is a bou[illegible]m that satisfies $|\varepsilon| \le \varepsilon_M (\varepsilon_M > 0)$.

5.2.2 [illegible]gence Analysis

[illegible]vergence of the designed nonlinear observer incorporating RBF neural net-[illegible]rk concept, the function $V(t) \in \mathbb{R}$ is selected as the Lyapunov function as following:

$$V = \tfrac{1}{2}p_1\tilde{x}_1^2 + \tfrac{1}{2}\tilde{x}_2^2 + \tfrac{1}{2}\tilde{x}_3^2 + \tfrac{1}{2}\tilde{x}_4^2 + \tfrac{1}{2}\tilde{W}^T\Gamma^{-1}\tilde{W} \tag{5.21}$$

where Γ^{-1} is the inverse matrix of Γ with positive definiteness property and p_1 is a positive constant in (5.13).

Theorem 5.1 *The estimation error of the SOC in the designed nonlinear observer incorporating RBF neural network concept in (5.16) is uniformly bounded provided that the observer gain* l_4 *satisfies*

$$l_4 > \frac{2p_1l_1^2}{a} - \frac{3(d_3+|l_2|)^2}{d_1} - \frac{3l_3^2}{d_2}. \tag{5.22}$$

Proof Since W^* is a constant weight vector, it can be deduced that $\dot{\tilde{W}} = \dot{\hat{W}}$. From (5.18)–(5.20), the derivative of $V(t)$ in (5.21) satisfies

$$\begin{aligned}\dot{V} &= p_1\tilde{x}_1\dot{\tilde{x}}_1 + \tilde{x}_2\dot{\tilde{x}}_2 + \tilde{x}_3\dot{\tilde{x}}_3 + \tilde{x}_4\dot{\tilde{x}}_4 + \tilde{W}^T\Gamma^{-1}\dot{\hat{W}} \\ &\le -ap_1\tilde{x}_1^2 + |l_1p_1\tilde{x}_1\tilde{x}_4| - d_1\tilde{x}_2^2 + M_1|\tilde{x}_2| + |d_3\tilde{x}_2\tilde{x}_4| \\ &\quad + |l_2\tilde{x}_2\tilde{x}_4| - d_2\tilde{x}_3^2 + M_2|\tilde{x}_3| + |l_3\tilde{x}_3\tilde{x}_4| - d_2\tilde{x}_4^2 \\ &\quad + M_3|\tilde{x}_4| - l_4\tilde{x}_4^2 + \varepsilon_M|\tilde{x}_4| - \tilde{W}^TK_wW.\end{aligned} \tag{5.23}$$

Based on the Lemma A.17 in [10], the following inequalities can be obtained:

$$\begin{aligned}&-ap_1\tilde{x}_1^2 - |l_1p_1\tilde{x}_1\tilde{x}_4| \le -\frac{ap_1}{2}\tilde{x}_1^2 + \frac{2p_1l_1^2\tilde{x}_4^2}{a} \\ &-d_1\tilde{x}_2^2 + M_1|\tilde{x}_2| + |d_3\tilde{x}_2\tilde{x}_4| + |l_2\tilde{x}_2\tilde{x}_4| \le -\frac{d_1}{3}\tilde{x}_2^2 + \frac{3M_1^2}{d_1} \\ &\quad + \frac{3(d_3+|l_2|)^2}{d_1}\tilde{x}_4^2 \\ &-d_2\tilde{x}_3^2 + M_2|\tilde{x}_3| + |l_3\tilde{x}_3\tilde{x}_4| \le -\frac{d_2}{3}\tilde{x}_3^2 + \frac{3M_2^2}{d_2} + \frac{3l_3^2}{d_2}\tilde{x}_4^2 \\ &-d_2\tilde{x}_4^2 + M_3|\tilde{x}_4| - l_4\tilde{x}_4^2 + \varepsilon_M|\tilde{x}_4| \le -l_4\tilde{x}_4^2 + \frac{(M_3+\varepsilon_M)^2}{d_2}.\end{aligned} \tag{5.24}$$

With the completion of squares, the following can be obtained:

$$\begin{aligned}-\tilde{W}^TK_wW &= -\tilde{W}^TK_w(\tilde{W}+W^*) \\ &\le -\frac{\|K_w\|\|\tilde{W}\|^2}{2} + \frac{\|K_w\|W_M^2}{2}.\end{aligned} \tag{5.25}$$

Based on (5.24) and (5.25), (5.23) is modified as:

$$\dot{V} \le -\frac{ap_1}{2}\tilde{x}_1^2 - \frac{d_1}{3}\tilde{x}_2^2 - \frac{d_2}{3}\tilde{x}_3^2 - \alpha\tilde{x}_4^2 - \frac{\|K_w\|\|\tilde{W}\|^2}{2} + M \tag{5.26}$$

with $\alpha = l_4 - \frac{2p_1l_1^2}{a} - \frac{3(d_3+|l_2|)^2}{d_1} - \frac{3l_3^2}{d_2}$, $M = \frac{3M_1^2}{d_1} + \frac{3M_2^2}{d_2} + \frac{(M_3+\varepsilon_M)^2}{d_2} + \frac{\|K_w\|W_M^2}{2}$. In reference to the analysis presented previously, $\frac{ap_1}{2}$, $\frac{d_1}{3}$, $\frac{d_2}{3}$, and $\frac{\|K_w\|}{2}$ are said to be positive. If the condition defined in (5.22) is satisfied, α can be guaranteed to be positive. Because of initial value $V(0)$ of the Lyapunov function is bounded, in reference to the boundedness analyzed in [11], the solution of the estimation error dynamics is uniformly bounded. Furthermore, the bound of estimation error of the SOC satisfies $\{|\tilde{x}_1(t)| \le \sqrt{\frac{M}{p}}\}$, where p is a number satisfying $0 < p < \frac{ap_1}{2}$. If p_1 is chosen to be very large, p could be selected to be large enough. The SOC estimation error can then be arbitrarily small. As a matter of fact, the SOC of the battery can be accurately estimated by the RBF neural network-based nonlinear observer.

5.3 Experiments and Simulations

The experiments have been performed with IFP36130155-36Ah lithium iron phosphate battery. The nominal capacity of the lithium-ion battery is 36 Ah, the nominal voltage is 3.7 V, and the cutoff voltage is 2.5 V. The battery and the test bench are depicted in Fig. 5.2. The test is performed at a temperature of 20°C, and a thermal governed chamber is used to ensure that the temperature of the battery fluctuates within a narrow range. The sampling rate of the current and voltage signals from the battery is set to 4 Hz. The experimental procedure is divided into two parts: an extraction procedure for all parameters in the propounded battery equivalent circuit model and an experiment to verify the performance of the proposed neural network-based nonlinear observer.

5.3.1 *Experiment for Parameter Extraction*

The process to extract the components in the circuit in the propounded battery model is based on the similar concepts depicted in [12]. The discharge capacity of the battery is 5% of the total capacity delivering a constant current of 12 A in steps, then the battery will be in rest for 20 min repeatedly until the SOC tends to zero. The battery's voltage curve is smoothened by incorporating a low-pass filter, which is illustrated in Fig. 5.3 with $SOC = 90\%$ and $SOC = 85\%$ for the voltage curve during the interval of rest. At the end of the period of rest, the OCV of the battery at the corresponding SOCs could be directly measured. Now, as depicted in Fig. 5.3,

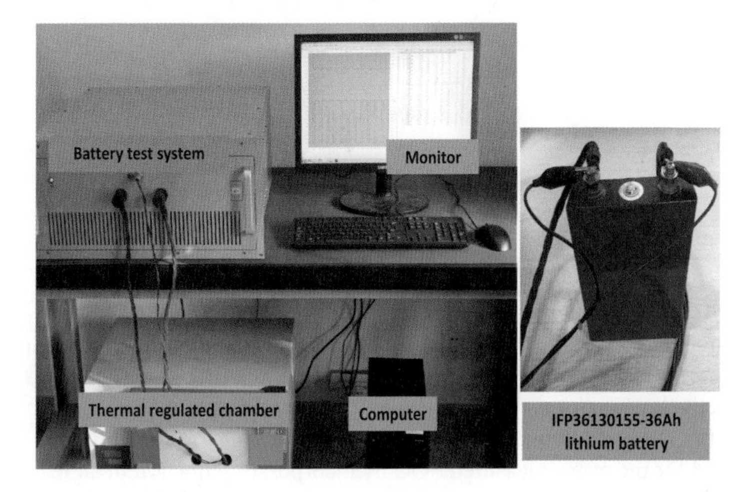

Fig. 5.2 Battery test bench

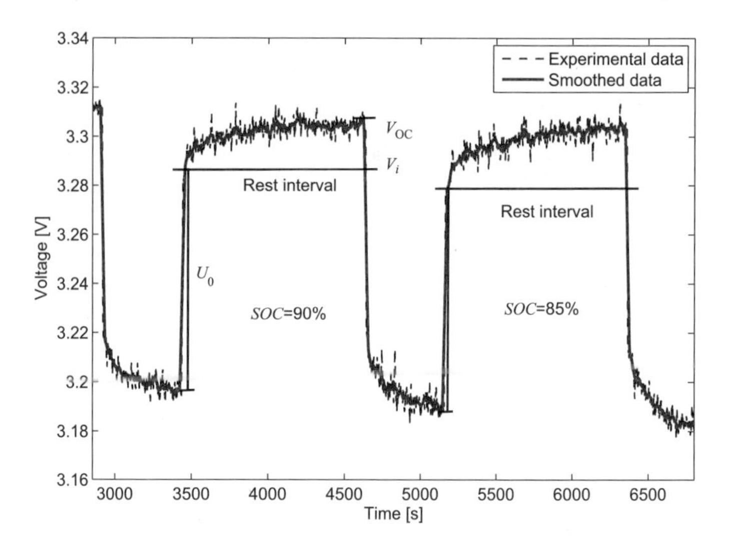

Fig. 5.3 Battery's terminal voltage response and its smoothed value

the resistance R_0 for the corresponding SOCs could be calculated as the ratio of the instantaneous voltage drop U_0 and the current I_B as:

$$R_0 = \frac{U_0}{I_B}. \tag{5.27}$$

The RC network parameters corresponding to the measured SOCs can be derived by fitting the *V-I* response curve to the following equations:

$$V_B = V_s(1 - e^{-\frac{t}{\tau_s}}) + V_f(1 - e^{-\frac{t}{\tau_f}}) + V_i \tag{5.28}$$

$$V_s = R_s I_B \tag{5.29}$$

$$V_f = R_f I_B = (\frac{V_{OC} - V_i}{I_B} - R_s) I_B \tag{5.30}$$

$$\tau_s = R_s C_s \ \tau_f = R_f C_f \tag{5.31}$$

where $V_B(t)$ is the terminal voltage of batteries; $I_B(t) = 12$ A is the battery's current before the rest interval; t represents the rest interval time duration; $V_s(t)$ and $V_f(t)$ are the short and long term transient voltage drops with the time constants τ_s and τ_f, respectively; $V_{OC}(t)$ is the OCV at the measured SOC and $V_i(t)$ is the battery's voltage after the instantaneous voltage rise. The variables R_s, C_s, R_f, and C_f can be obtained for the every measured SOC by incorporating the nonlinear least square method concept based on (5.28)–(5.31), where $V_B(t)$, $V_{OC}(t)$, $V_i(t)$, t and $I_B(t)$ are the inputs; τ_s, τ_f, $V_s(t)$ and $V_f(t)$ are the intermediate variables. Then, with a trade-off among accuracy and complexity, empirical equations composed of exponential

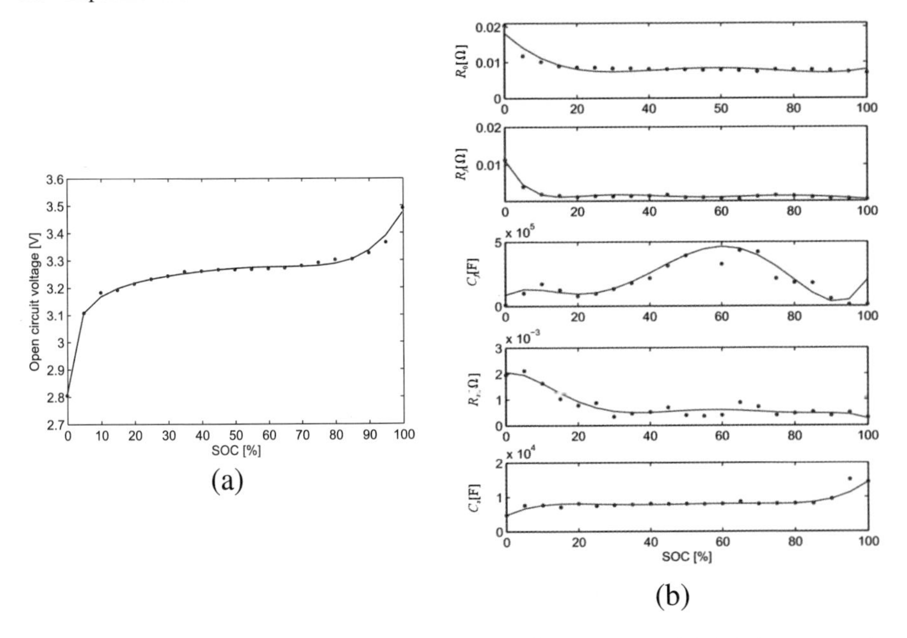

Fig. 5.4 **a** Open circuit voltage versus SOC of the battery, **b** Parameters and their fitting curves of the battery equivalent circuit model

functions and polynomials are used to fit these curves. The nonlinear relationship between the OCV and the SOC of the battery, and the curves of the parameters (R_0, R_f, C_f, R_s and C_s) in the battery equivalent circuit versus the SOC are illustrated in Fig. 5.4a, b, respectively. The red dots represent the experimental data points, and the blue lines represent the curves fitting. The comparison of the simulation result of the battery equivalent circuit model and the experimental data is shown in Fig. 5.5, demonstrating that the equivalent circuit model matches the real battery dynamics well.

5.3.2 Experiment for SOC Estimation

The number of nodes in the RBF neural network is taken as 7. The parameters in the neural network's weight vector's adaptive law are $\Gamma = \mathrm{diag}\{1, 1, 1, 1, 1, 1, 1\}$ and $K_w = \mathrm{diag}[0.01, 0.01, 0.01, 0.01, 0.01, 0.01, 0.01, 0.01]$. As shown in Fig. 5.6, a pulse current similar to [13] is used as the battery's discharging current. It has an amplitude of 12 A, a period of 1672 s, and a duty ratio of 30%. The observer's SOC is initially set to 50%. The RBF neural network's estimated uncertain term is shown in Fig. 5.7. Because the RBF neural network-based nonlinear observer design assumes the derivate of the current to be 0, there might be a large model bias for the step-in pulse current profile. The RBF neural network estimates the model bias, which is

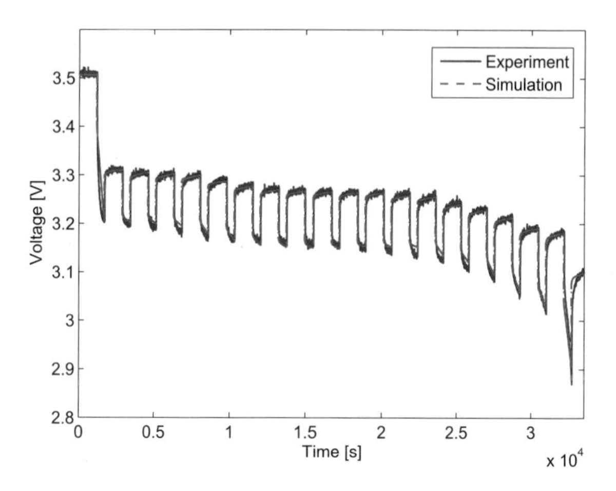

Fig. 5.5 Simulation result and experimental data of the battery's terminal voltage

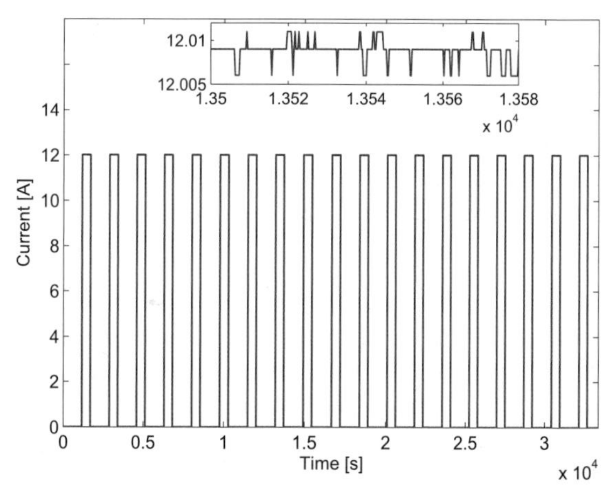

Fig. 5.6 Current profile of the battery

depicted in Fig. 5.7 as large spikes when the current profile steps-up and down. In Fig. 5.8, the curves of the actual battery's SOC, the SOC estimation by the EKF, and the estimated SOC utilizing the proposed nonlinear observer based on neural network concept are compared. The actual SOC is calculated by combining the ampere-hour counting method described in [13–15] with the initial SOC calculated by measuring the battery's initial OCV, with the current accurately measured by a power analyzer. The RMS errors in SOC in the designed RBF neural network based observer and in the EKF are 1.21% and 3.06%, respectively, as shown in Fig. 5.9 and Table 5.1. The designed nonlinear observer outperforms the EKF in estimation accuracy.

The corresponding SOC estimation errors are depicted in Fig. 5.10 for different initial estimated SOC values: $SOC(0) = 50\%$, $SOC(0) = 70\%$, and $SOC(0) = 90\%$. As shown in Table 5.2, the average SOC estimation RMS error of the designed nonlinear observer based on the RBF neural network is 1.1233% with initial estimated SOC values of 50%, 70%, and 90%, demonstrating its good SOC estimation performance.

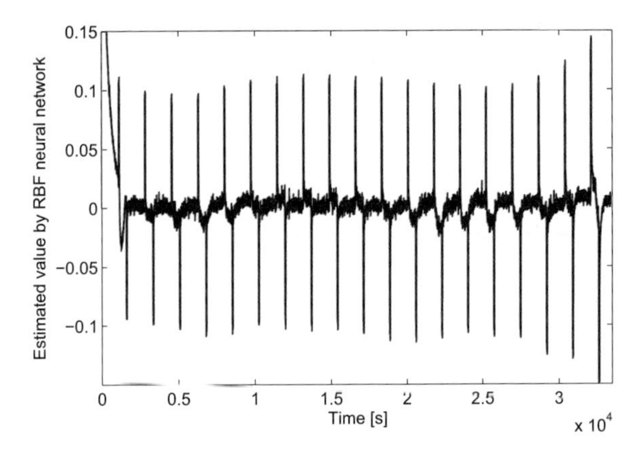

Fig. 5.7 Estimated uncertain term by the RBF neural network

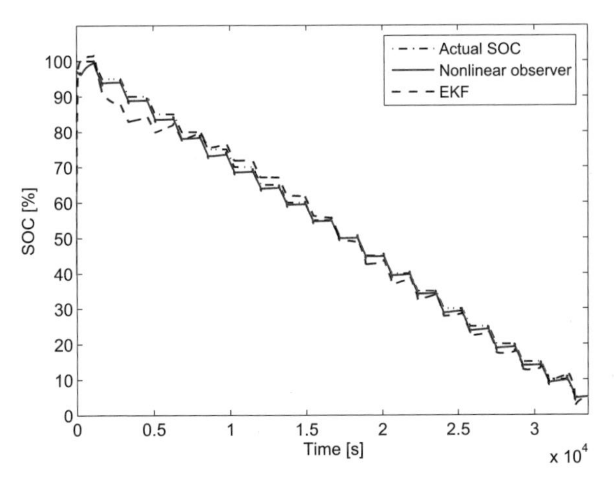

Fig. 5.8 Comparison of the actual SOC, the SOC estimated by EKF, and the SOC estimated by nonlinear observer

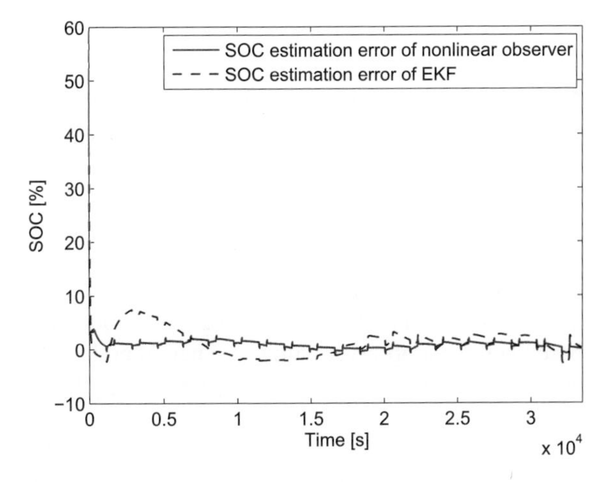

Fig. 5.9 Comparison of SOC estimation errors by EKF and nonlinear observer

Table 5.1 Statistics of SOC estimation errors with pulse current

	Mean (%)	RMS (%)	Variance
Nonlinear observer	0.88	1.21	6.9189×10^{-5}
EKF	1.16	3.06	8.0197×10^{-4}

Table 5.2 Statistical results of SOC estimation errors with different initial estimated SOC values

Initial estimated SOC (%)	Mean (%)	RMS (%)	Variance
50	0.88	1.21	6.9189×10^{-5}
70	0.86	1.12	5.056×10^{-5}
90	0.82	1.04	4.1265×10^{-5}

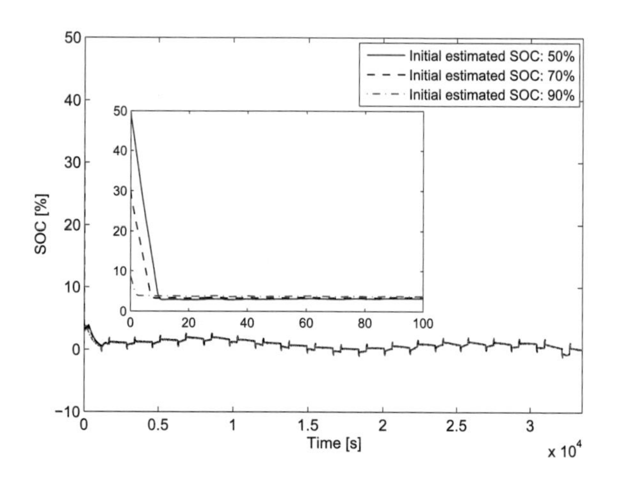

Fig. 5.10 SOC estimation errors of the nonlinear observers with different initial estimated SOC values

Remark 5.3 The effect of neuron count on SOC estimation performance and computational burden of the designed nonlinear observer incorporating neural network concepts is investigated through comparison experiments and simulations. In particular, we use the experimental data of the battery's current and terminal voltage to implement RBF neural network-based nonlinear observers with two, five, seven, eleven, and fifteen neural nodes in Simulink/MATLAB. Figure 5.11a depicts the comparison results in terms of SOC estimation performance. Figure 5.11b depicts the time elapsed of Simulink programmes for solving nonlinear observer problems with different neuron numbers. According to the comparison results, the neural network-based nonlinear observer with seven neural nodes used in our experiments produces good performance while being computationally light.

When the battery is functioned with a complex current profile, simulations in Simulink/MATLAB utilizing experimental data are also performed to validate the envisaged SOC estimation algorithm. The FUDS is a typical driving cycle that is

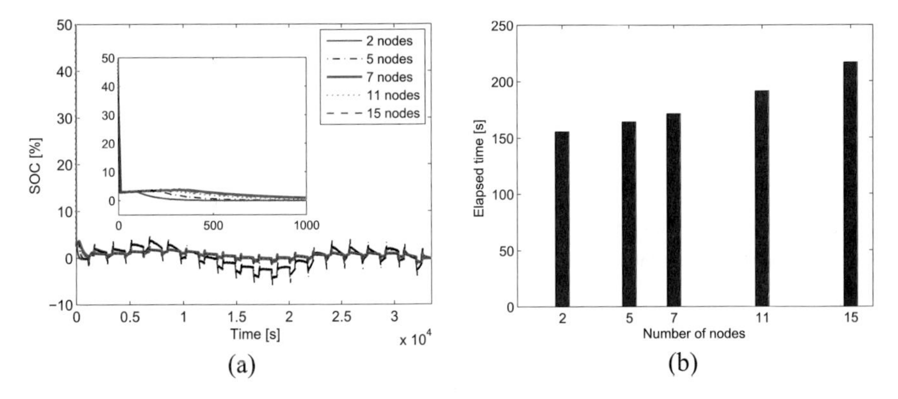

Fig. 5.11 **a** SOC estimation errors of the nonlinear observers with different numbers of neural nodes, **b** elapsed times of the Simulink programs of solving the nonlinear observers with different numbers of neural nodes

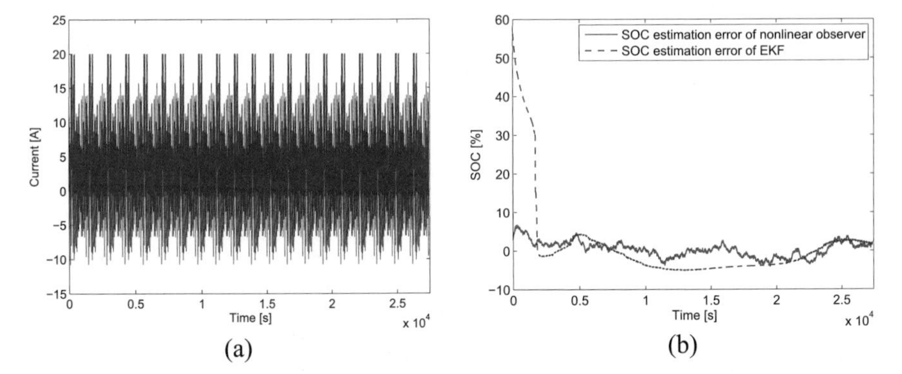

Fig. 5.12 **a** Current profile sampled during 20 consecutive FUDS cycles, **b** comparison of SOC estimation errors of EKF and nonlinear observer

frequently used to evaluate SOC estimation algorithms [16], where the variable charge/discharge regimes indicate the best simulation available of an electric vehicle's actual power demand and regenerative power, respectively. The FUDS test cycle is repeated 20 times in order to charge and discharge the battery using the current profile shown in Fig. 5.12a. The observer's initial SOC is set to 50%. The RMS error in SOC estimation by the RBF neural network based nonlinear observer under 20 consecutive FUDS cycles is 2.23%, as shown in Fig. 5.12b and Table 5.3. The proposed observer has a faster convergence speed and higher precision than the EKF. At last, another two tests composed of 20 consecutive FUDS cycles are carried out to verify the SOC estimation performance of the designed RBF neural network based nonlinear observer, in which the maximum current magnitudes are 36 A and 50 A, respectively. The statistical results of the SOC estimation errors are shown in Table 5.4. The average SOC estimation RMS error of the designed nonlin-

Table 5.3 Statistical analysis of SOC estimation errors under 20 consecutive FUDS cycles

	Mean (%)	RMS (%)	Variance
Nonlinear observer	0.35	2.23	4.8658×10^{-4}
EKF	1.16	10.81	0.0115

Table 5.4 SOC estimation errors for FUDS cycles with different maximum current magnitudes

Maximum current magnitude (A)	RMS (%)	Variance
20	2.23	4.8658×10^{-4}
36	2.37	4.3397×10^{-4}
50	2.48	3.9458×10^{-4}

ear observer is 2.36% for FUDS cycles with maximum current magnitudes of 20 A, 36 A, and 50 A, which indicates the satisfactory SOC estimation performance of the RBF neural network based nonlinear observer.

References

1. M. Chen, G. Rincon-Mora, Accurate electrical battery model capable of predicting runtime and I–V performance. IEEE Trans. Energy Convers. **21**(2), 504–511 (2006)
2. R. Van Gils, D. Danilov, P. Notten, M. Speetjens, H. Nijmeijer, Battery thermal management by boiling heat-transfer. Energy Convers. Manag. **79**, 9–17 (2014)
3. Q. Ouyang, J. Chen, F. Wang, H. Su, Nonlinear observer design for the state of charge of lithium-ion batteries. IFAC Proc. **47**(3), 2794–2799 (2014)
4. T. Kim, W. Qiao, A hybrid battery model capable of capturing dynamic circuit characteristics and nonlinear capacity effects. IEEE Trans. Energy Convers. **26**(4), 1172–1180 (2011)
5. M. Sitterly, L.Y. Wang, G.G. Yin, C. Wang, Enhanced identification of battery models for real-time battery management. IEEE Trans. Sustain. Energy **2**(3), 300–308 (2011)
6. M. Partovibakhsh, G. Liu, An adaptive unscented kalman filtering approach for online estimation of model parameters and state-of-charge of lithium-ion batteries for autonomous mobile robots. IEEE Trans. Control. Syst. Technol. **23**(1), 357–363 (2015)
7. M. Chen, S.S. Ge, B.V.E. How, Y.S. Choo, Robust adaptive position mooring control for marine vessels. IEEE Trans. Control. Syst. Technol. **21**(2), 395–409 (2013)
8. G. Sun, X. Ren, D. Li, Neural active disturbance rejection output control of multimotor servomechanism. IEEE Trans. Control. Syst. Technol. **23**(2), 746–753 (2015)
9. J. Park, I.W. Sandberg, Universal approximation using radial-basis-function networks. Neural Comput. **3**(2), 246–257 (1991)
10. M.S. De Queiroz, D.M. Dawson, S.P. Nagarkatti, F. Zhang, *Lyapunov-Based Control of Mechanical Systems* (Springer Science & Business Media, 2000)
11. H.K. Khalil, *Nonlinear Systems* (Prentice Hall, Englewood Cliffs, NJ, USA, 2002)
12. S. Abu-Sharkh, D. Doerffel, Rapid test and non-linear model characterisation of solid-state lithium-ion batteries. J. Power Sources **130**(1–2), 266–274 (2004)
13. M. Gholizadeh, F.R. Salmasi, Estimation of state of charge, unknown nonlinearities, and state of health of a lithium-ion battery based on a comprehensive unobservable model. IEEE Trans. Ind. Electron. **61**(3), 1335–1344 (2013)

14. S. Lee, J. Kim, J. Lee, B.H. Cho, State-of-charge and capacity estimation of lithium-ion battery using a new open-circuit voltage versus state-of-charge. J. Power Sources **185**(2), 1367–1373 (2008)
15. S. Dey, B. Ayalew, P. Pisu, Nonlinear robust observers for state-of-charge estimation of lithium-ion cells based on a reduced electrochemical model. IEEE Trans. Control. Syst. Technol. **23**(5), 1935–1942 (2015)
16. H. He, R. Xiong, X. Zhang, F. Sun, J. Fan, State-of-charge estimation of the lithium-ion battery using an adaptive extended kalman filter based on an improved thevenin model. IEEE Trans. Veh. Technol. **60**(4), 1461–1469 (2011)

Chapter 6
Active Cell-to-Cell Equalization Control

6.1 Cell Equalizing System Model

As shown in Fig. 6.1, an n-modular serially connected lithium-ion battery pack requires $n-1$ bidirectional modified Cûk converters as cell equalizers. Through the connected converters, energy can be transferred from one cell to another. The cell model and the converter model are presented in this section. The cell equalizing system model is then proposed for an n-modular battery pack.

6.1.1 Battery Cell Model

As shown in Fig. 6.2, an intuitive and comprehensive equivalent circuit model [1] is used to characterize the i-th ($1 \le i \le n$) cell's characteristics. The impacts of the cell's cycle number and temperature on the model's equivalent circuit are ignored within certain error tolerances. The left subcircuit simulates the i-th cell's SOC, and the right subcircuit reflects its transient voltage-current response, where C_{b_i} signifies the cell's capacity, R_{sd_i} defines the self-discharge resistor, R_{0_i} is used to characterize the charge and discharge energy losses, and RC networks (R_{s_i}, C_{s_i}) and (R_{f_i}, C_{f_i}) indicate the short-term and long-term transient responses.

For the i-th cell, referring to [2], its dynamics can be derived as:

$$
\begin{aligned}
\dot{V}_{SOC_i} &= -\frac{\eta_0}{R_{sd_i} C_{b_i}} V_{SOC_i} - \frac{\eta_0}{C_{b_i}} I_{B_i} \\
\dot{V}_{s_i} &= -\frac{1}{R_{s_i} C_{s_i}} V_{s_i} + \frac{1}{C_{s_i}} I_{B_i} \\
\dot{V}_{f_i} &= -\frac{1}{R_{f_i} C_{f_i}} V_{f_i} + \frac{1}{C_{f_i}} I_{B_i} \\
V_{B_i} &= V_{OC_i} - R_{0_i} I_{B_i} - V_{f_i} - V_{s_i}
\end{aligned}
\tag{6.1}
$$

J. Chen et al., *Equalization Control for Lithium-Ion Batteries*,
https://doi.org/10.1007/978-981-99-0220-0_6

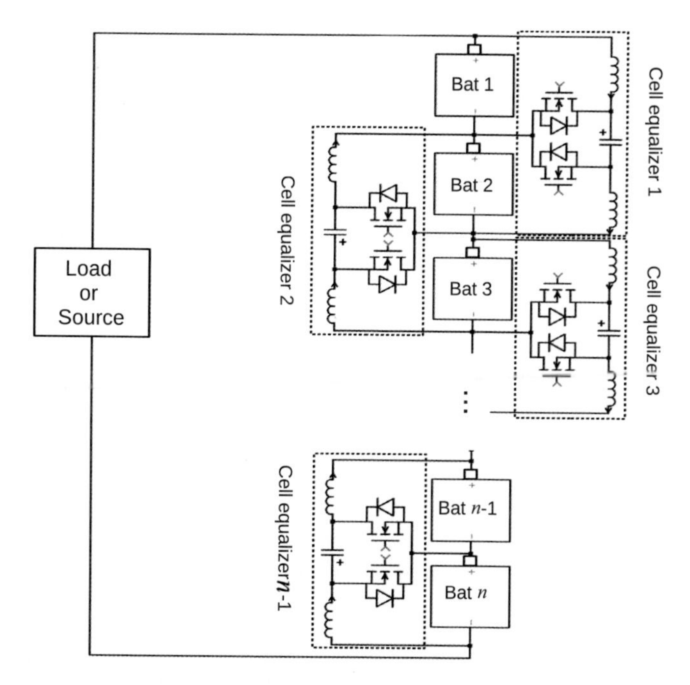

Fig. 6.1 Diagram of a serially connected battery pack with cell equalizers

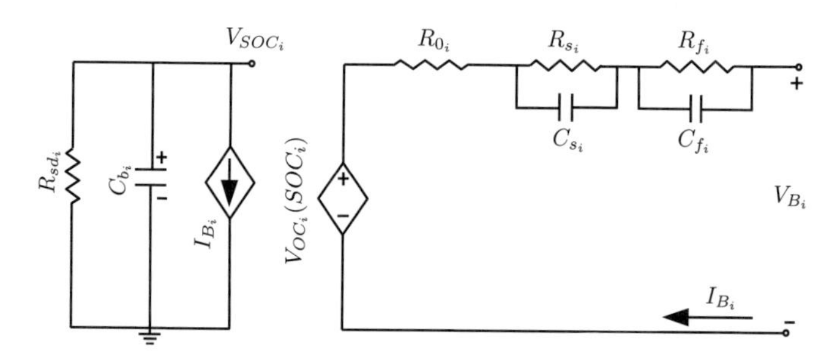

Fig. 6.2 Equivalent circuit model of the battery

where the variables V_{SOC_i}, V_{s_i}, and V_{f_i} represent the voltages across the capacitors C_{b_i}, C_{s_i}, and C_{f_i}, respectively; V_{B_i} and V_{OC_i} are the i-th cell's terminal voltage and OCV, respectively; I_{B_i} is the i-th cell's current, which is positive/negative when the cell is in the discharging/charging mode; η_0 indicates the Coulombic efficiency, which satisfies $\eta_0 = 1$ for discharging, and $\eta_0 = \eta_c \leq 1$ for charging with η_c equal to the ratio between the cell's discharging and charging capacities [3]. The variable V_{SOC_i} defines the i-th cell's SOC quantitatively, with 0–1 V of V_{SOC_i} corresponding to 0–100% of the SOC. The mapping from the i-th cell's SOC to its OCV is expressed as:

$$V_{OC_i} = k_{1_i} + k_{2_i} SOC_i + \Delta V_{OC_i} \tag{6.2}$$

where k_{1_i} and k_{2_i} are positive constants, SOC_i represents the i-th cell's SOC, ΔVoc_i includes the nonlinear part and the unknown disturbance of the relationship between the i-th cell's OCV and SOC. Because the value of V_{SOC_i} equals to the i-th cell's SOC, SOC_i could be used to replace V_{SOC_i} in the battery model. The derivate of the cell's current I_{B_i} can be assumed to be negligible due to the fast-sampling rate [2]. The derivative of the terminal voltage is calculated using (6.1) and (6.2) as follows:

$$\begin{aligned}\dot{V}_{B_i} = & \frac{k_{2_i} SOC_i}{R_{f_i} C_{f_i}} - \frac{k_{2_i} \eta_0 SOC_i}{R_{sd_i} C_{b_i}} - \frac{V_{B_i}}{R_{f_i} C_{f_i}} + \frac{V_{s_i}}{R_{s_i} C_{s_i}} \\ & - \frac{V_{s_i}}{R_{f_i} C_{f_i}} - \left(\frac{1}{C_{s_i}} + \frac{1}{C_{f_i}} + \frac{R_{0_i}}{R_{f_i} C_{f_i}} + \frac{k_{2_i} \eta_0}{C_{b_i}} \right) I_{B_i} \\ & + \frac{k_{1_i}}{R_{f_i} C_{f_i}} + \varphi_i(SOC_\iota, I_{B_i})\end{aligned} \tag{6.3}$$

where $\varphi_i(SOC_i, I_{B_i})$ is the nonlinear and unknown term of the terminal voltage's dynamics. Based on (6.1)–(6.3), the i-th cell's model can be rewritten as the following state space realization

$$\begin{aligned}\dot{x}_i &= A_i x_i + b_i \varepsilon_i + e_i + m_i \varphi_i(x_i, \varepsilon_i) \\ y_i &= c_i x_i\end{aligned} \tag{6.4}$$

where $x_i \triangleq [SOC_i, V_{s_i}, V_{f_i}, V_{B_i}]^T \in \mathbb{R}^4$ is the state vector; $y_i \triangleq V_B \in \mathbb{R}$ is the output; $\varepsilon_i \triangleq I_B \in \mathbb{R}$ is treated as a measurable disturbance; $A_i \in \mathbb{R}^{4\times4}$, $b_i \in \mathbb{R}^4$, $e_i \in \mathbb{R}^4$, $m_i \in \mathbb{R}^4$, and $c_i \in \mathbb{R}^{1\times4}$ are defined as

$$A_i = \begin{bmatrix} -a_{i_1} & 0 & 0 & 0 \\ 0 & -a_{i_2} & 0 & 0 \\ 0 & 0 & -a_{i_3} & 0 \\ k_{2_i}(a_{i_3} - a_{i_1}) & a_{i_2} - a_{i_3} & 0 & -a_{i_3} \end{bmatrix}$$

$$b_i = \left[-\frac{\eta_0}{C_{b_i}} \ \frac{1}{C_{s_i}} \ \frac{1}{C_{f_i}} \ -\frac{1}{C_{s_i}} - \frac{1}{C_{f_i}} - R_{0_i} a_{i_3} - \frac{k_{2_i} \eta_0}{C_{b_i}} \right]^T$$

$$e_i = \left[0\ 0\ 0\ k_{1_i} a_{i_3} \right]^T, m_i = \left[0\ 0\ 0\ 1 \right]^T$$

$$c_i = \left[0\ 0\ 0\ 1 \right]$$

with $a_{i_1} = \frac{\eta_0}{C_{b_i} R_{sd_i}}$, $a_{i_2} = \frac{1}{C_{s_i} R_{s_i}}$, and $a_{i_3} = \frac{1}{C_{f_i} R_{f_i}}$. For the battery system (6.4), $\varphi_i(x_i, \varepsilon_i)$ can be assumed to be bounded and satisfies

$$|\varphi_i(x_i, \varepsilon_i)| \leq \varphi_{iM} \tag{6.5}$$

where φ_{i_M} is a positive constant for $1 \leq i \leq n$.

6.1.2 Bidirectional Modified Cûk Converter Model

The battery equalizer here employs the Cûk converter, which has three operating states. Refer the bidirectional modified Cûk converter in the commercial equalizers above for the specific working states.

The i-th ($1 \leq i \leq n-1$) converter could be incorporated to transfer the energy from i-th to ($i+1$)-th cells in bidirectional mode following structural symmetricity. As depicted in Fig. 6.1, the i-th bidirectional modified Cûk converter comprises of uncoupled inductors L_{i1} and L_{i2}, an energy transferring capacitor C_i, and the MOSFETs Q_{i1} and Q_{i2} with body diodes d_{i1} and d_{i2} [4]. MOSFETs Q_{i1} and Q_{i2} are regulated by PWM signals to be on or off as the cell balancing control switches, but they cannot be turned on simultaneously [4]. The converter is designed to operate in DICM to enhance the average energy transfer efficiency over CICM [5]. The control variable for the inductor currents that transfer energy from the i-th/($i+1$)-th cell to the ($i+1$)-th/i-th cell is the PWM duty cycle applied to MOSFET Q_{i1}/Q_{i2}. In reference to [6], the average inductor currents are formulated when the charge is transferred from the i-th/($i+1$)-th cell to the ($i+1$)-th/i-th cell as follows:

$$
\begin{aligned}
I_{L_{i1}} &= \begin{cases} f_{i1}(D_{i1}), & \text{energy from cell } i \text{ to cell } i+1 \\ p_i' f_{i2}(D_{i2}), & \text{energy from cell } i+1 \text{ to cell } i \end{cases} \\
I_{L_{i2}} &= \begin{cases} p_i f_{i1}(D_{i1}), & \text{energy from cell } i \text{ to cell } i+1 \\ f_{i2}(D_{i2}), & \text{energy from cell } i+1 \text{ to cell } i \end{cases}
\end{aligned} \tag{6.6}
$$

with

$$
\begin{aligned}
f_{i1}(D_{i1}) &= \frac{1}{2}\frac{T_s V_{B_i} V_{C_i} D_{i1}^2}{L_{i1}(V_{C_i} - V_{B_i})} \\
f_{i2}(D_{i2}) &= -\frac{1}{2}\frac{T_s V_{B_{i+1}} V_{C_i} D_{i2}^2}{L_{i2}(V_{C_i} - V_{B_{i+1}})}
\end{aligned} \tag{6.7}
$$

where D_{i1} and D_{i2} represent the duty cycles of the PWM signals applied to MOSFET Q_{i1} and Q_{i2}, respectively; T_s is the switching period; $I_{L_{i1}}$, $I_{L_{i2}}$, and V_{C_i} indicate the average inductor currents and the average capacitor voltage, which are constants for a particular duty cycle D_{i1} or D_{i2} [4]; V_{B_i} and $V_{B_{i+1}}$ are the terminal voltages of the i-th and ($i+1$)-th cells, respectively; p_i and p_i' are the energy transfer efficiencies satisfying $0 \leq p_i \leq 1$ and $0 \leq p_i' \leq 1$. The direction of current coming out of the i-th cell and going into the ($i+1$)-th cell are defined as the reference direction of $I_{L_{i1}}$ and $I_{L_{i2}}$, respectively. In reference to [4], V_{C_i} could be obtained by approximating $V_{C_i} \approx V_{B_i} + V_{B_{i+1}}$. The negative sign of $f_{i2}(D_{i2})$ in (6.7) indicates that it is pointing in the reverse direction as the pre-defined reference direction. When energy is transferred from the i-th/($i+1$)-th cell to the ($i+1$)-th/i-th cell, the average inductor current $I_{L_{i1}}/I_{L_{i2}}$ is referred to as the controlled equalizing current. According to (6.6) and (6.7), the corresponding duty cycles for a typical controlled equalizing current $I_{L_{i1}}/I_{L_{i2}}$ are as follows:

$$D_{i1} = \begin{cases} \sqrt{\frac{2L_{i1}(V_{C_i}-V_{B_i})I_{L_{i1}}}{T_s V_{B_i} V_{C_i}}}, & \text{energy from cells } i \text{ to } i{+}1 \\ 0, & \text{energy from cells } i{+}1 \text{ to } i \end{cases}$$
$$D_{i2} = \begin{cases} 0, & \text{energy from cells } i \text{ to } i{+}1 \\ \sqrt{-\frac{2L_{i2}(V_{C_i}-V_{B_{i+1}})I_{L_{i2}}}{T_s V_{B_{i+1}} V_{C_i}}}, & \text{energy from cells } i{+}1 \text{ to } i. \end{cases} \tag{6.8}$$

The following work will concentrate on the design of controlled equalizing currents, and their corresponding duty cycles could be determined using (6.8).

6.1.3 Cell Equalizing System Model

The i-th $(1 \leq i \leq n)$ cell's current $I_{B_i}(k)$ can be calculated with the assumption that the external current is constant throughout every sampling period for the serially connected battery pack as follows:

$$I_{B_i}(k) = I_s(k) + I_{eq_i}(k) \tag{6.9}$$

where $I_s(k)$ is the battery pack's exogenous current to the load or from the charger, that is identical for all cells due to the series connection; and $I_{eq_i}(k)$ is the equalizing current for the i-th cell. The equalizing current of the i-th $(2 \leq I \leq n-1)$ cell is composed of currents to or from its adjacent cells via its connected $(i-1)$-th and i-th converters, as shown in Fig. 6.1. The balancing current is only received by the first and n-th cells from the second and $(n-1)$-th converters. As a result, the cells' equalizing currents can be expressed as:

$$\begin{aligned} I_{eq_1}(k) &= I_{L_{11}}(k) \\ I_{eq_i}(k) &= -I_{L_{(i-1)2}}(k) + I_{L_{i1}}(k) \;\; (2 \leq i \leq n-1) \\ I_{eq_n}(k) &= -I_{L_{(n-1)2}}(k). \end{aligned} \tag{6.10}$$

Since MOSFETs Q_{i1} and Q_{i2} $(1 \leq i \leq n-1)$ cannot be turned on simultaneously, $D_{i1}(k)D_{i2}(k) = 0$. Taking the model biases of the controlled equalizing currents into account, (6.6) can be rewritten as follows:

$$\begin{aligned} I_{L_{i1}}(k) &= \gamma_i(f_{i1}(D_{i1}(k)) + w_{i1}(k)) + \gamma_i' p_i'(f_{i2}(D_{i2}(k)) + w_{i2}(k)) \\ I_{L_{i2}}(k) &= \gamma_i p_i(f_{i1}(D_{i1}(k)) + w_{i1}(k)) + \gamma_i'(f_{i2}(D_{i2}(k)) + w_{i2}(k)) \end{aligned} \tag{6.11}$$

with

$$\gamma_i = \begin{cases} 1, & D_{i1}(k) \neq 0 \\ 0, & D_{i1}(k) = 0 \end{cases}, \; \gamma_i' = \begin{cases} 0, & D_{i2}(k) = 0 \\ 1, & D_{i2}(k) \neq 0 \end{cases}$$

for $1 \leq i \leq n-1$, where $w_{i1}(k)$ and $w_{i2}(k)$ represent the unknown model biases and disturbances of the controlled equalizing current through the i-th converter. By substituting (6.11) into (6.10), the following expressions can be obtained:

$$\begin{aligned}
I_{eq_1}(k) &= \gamma_1(f_{11}(D_{11}(k)) + w_{11}(k)) + \gamma_1' p_1'(f_{12}(D_{12}(k)) + w_{12}(k)) \\
I_{eq_i}(k) &= -\gamma_{i-1} p_{i-1}(f_{(i-1)1}(D_{(i-1)1}(k)) + w_{(i-1)1}(k)) - \gamma_{i-1}' \\
&\quad (f_{(i-1)2}(D_{(i-1)2}(k)) + w_{(i-1)2}(k)) + \gamma_i(f_{i1}(D_{i1}(k)) \\
&\quad + w_{i1}(k)) + \gamma_i' p_i'(f_{i2}(D_{i2}(k)) + w_{i2}(k)) \qquad (2 \leq i \leq n-1) \\
I_{eq_n}(k) &= -\gamma_{n-1} p_{n-1}(f_{(n-1)1}(D_{(n-1)1}(k)) + w_{(n-1)1}(k)) \\
&\quad - \gamma_{n-1}'(f_{(n-1)2}(D_{(n-1)2}(k)) + w_{(n-1)2}(k)).
\end{aligned} \tag{6.12}$$

In practical applications, the capacities of the cells are selected as same, defined as C_b for serially connected battery pack. If we neglect the self-discharging characteristics in the battery model (6.1), the change in i-th $(1 \leq i \leq n)$ cell's SOC in a particular sampling period can be calculated as follows:

$$SOC_i(k+1) = SOC_i(k) - dI_{B_i}(k) \tag{6.13}$$

with $d = \eta_0 T / C_b$ as an auxiliary variable, where T is the sampling period, d is a very small constant, because $C_b = 3600\times$ Ampere-hour capacity [1]. In reference to (6.9), (6.12), and (6.13), the cell equalizing system model for the n-modular serially connected battery pack is as follows:

$$z(k{+}1) = z(k){+}dB_1(k)(u_1(k){+}w_1(k)){+}dB_2(k)(u_2(k){+}w_2(k)){-}b(k) \tag{6.14}$$

where $z(k) \triangleq [x_{1_1}(k), \ldots, x_{n_1}(k)]^T \in \mathbb{R}^n$ indicates the SOC vector in the serially connected battery pack; $u_1(k) \triangleq [f_{11}(D_{11}(k)), \ldots, f_{(n-1)1}(D_{(n-1)1}(k))]^T \in \mathbb{R}^{n-1}$ and $u_2(k) \triangleq [f_{12}(D_{12}(k)), \ldots, f_{(n-1)2}(D_{(n-1)2}(k))]^T \in \mathbb{R}^{n-1}$ are the controlled equalizing currents, respectively; $w_1(k) \triangleq [w_{11}(k), \ldots, w_{(n-1)1}(k)]^T \in \mathbb{R}^{n-1}$ and $w_2(k) \triangleq [w_{12}(k), \ldots, w_{(n-1)2}(k)]^T \in \mathbb{R}^{n-1}$ are the unknown model biases; $B_1(k) \in \mathbb{R}^{n\times(n-1)}$, $B_2(k) \in \mathbb{R}^{n\times(n-1)}$, and $b(k) \in \mathbb{R}^n$ are defined as:

$$B_1(k) = \begin{bmatrix} -\gamma_1 & 0 & \cdots & 0 \\ \gamma_1 p_1 & -\gamma_2 & \cdots & 0 \\ \cdots & \cdots & \cdots & \cdots \\ 0 & 0 & \cdots & \gamma_{n-1} p_{n-1} \end{bmatrix}$$

$$B_2(k) = \begin{bmatrix} -\gamma_1' p_1' & 0 & \cdots & 0 \\ \gamma_1' & -\gamma_2' p_2' & \cdots & 0 \\ \cdots & \cdots & \cdots & \cdots \\ 0 & 0 & \cdots & \gamma_{n-1}' \end{bmatrix}$$

$$b(k) = \left[dI_s(k), \cdots, dI_s(k)\right]^T.$$

6.2 Objective and Constraints of the Cell Equalizing Process

The model of the cell equalizing system was proposed in the preceding section. In this section, we first present the goal of the cell equalizing control design before listing the cell balancing constraints.

6.2.1 *Cell Equalizing Objective*

The primary objective of cell equalization is for the SOC differences between serially connected cells to converge to a tolerance, which can be expressed as:

$$|z_i(k) - \bar{z}(k)| \leq \epsilon, \ \forall k \geq \tau \tag{6.15}$$

with $\bar{z}(k) = \frac{1}{n}\sum_{i=1}^{n} z_i(k)$ for all initial values $z_i(0)$ $(1 \leq i \leq n)$, where $z_i(k)$ is the i-th cell's SOC, $\bar{z}(k)$ denotes the cells' average SOC, ϵ is the cells' maximum tolerable SOC difference, and τ is the cell equalizing time. As MOSFETs Q_{i1} and Q_{i2} $(1 \leq i \leq n-1)$ cannot be turned on simultaneously, there are $n-1$ redundant control variables in (6.14), which are set as zero. From (6.7), it yields that $u_{i1}(k) \geq 0$ and $u_{i2}(k) \leq 0$, where $u_{i1}(k)$ and $u_{i2}(k)$ are the i-th elements of $u_1(k)$ and $u_2(k)$, respectively. Then, we can use a new input $u(k) \in \mathbb{R}^{n-1}$ to replace $u_1(k)$ and $u_2(k)$ for the sake of redundant input reduction as follows:

$$\begin{cases} u_{i1}(k) = u_i(k), \gamma_i = 1, \ u_i(k) \geq 0 \\ u_{i1}(k) = 0, \gamma_i = 0, \qquad u_i(k) < 0 \\ u_{i2}(k) = 0, \gamma_i' = 0, \qquad u_i(k) \geq 0 \\ u_{i2}(k) = u_i(k), \gamma_i' = 1, \ u_i(k) < 0 \end{cases} \tag{6.16}$$

where $u_i(k)$ is the i-th element of $u(k)$. If we remove the redundant control variables, the model of cell equalizing system (6.14) can be written in the simplified form as follows:

$$z(k+1) = z(k) + d\bar{C}\,(u(k) + w(k)) - b(k) \tag{6.17}$$

where $\bar{C} \in \mathbb{R}^{n\times(n-1)}$ is the corresponding coefficient matrix, the unknown model bias $w(k) = [\,w_1(k), \ldots, w_{n-1}(k)]^T \in \mathbb{R}^{n-1}$ is bounded and satisfies

$$|w_i(k)| \leq w_M \tag{6.18}$$

for $1 \leq i \leq n-1$ with w_M a small constant.

Example For a four-modular serially connected battery pack with controlled equalizing currents, $u_1(k) > 0$, $u_2(k) < 0$, and $u_3(k) > 0$ are designed based on the technique proposed in (6.16), it can be obtained that $u_{12}(k)=0$, $\gamma_1'=0$, $u_{21}(k)=0$, $\gamma_2=0$, $u_{32}(k)=0$, $\gamma_3'=0$, $u_{11}(k)=u_1(k)$, $\gamma_1=1$, $u_{22}(k)=u_2(k)$, $\gamma_2'=1$, $u_{31}(k)=u_3(k)$, and $\gamma_3=1$, respectively. By substituting them into (6.14), the coefficient matrix $\bar{C}$ in (6.17) is

$$\bar{C}=B_1(k)+B_2(k)=\begin{bmatrix}-1 & 0 & 0\\ p_1 & -p_2' & 0\\ 0 & 1 & -1\\ 0 & 0 & p_3\end{bmatrix}. \tag{6.19}$$

6.2.2 *Cell Equalizing Constraints*

DICM operation constraints: The controlled equalizing current of the i-th ($1 \le I \le n-1$) converter must satisfy the boundary condition to ensure the bidirectional modified Cûk converters operate in DICM. The boundary condition can be represented as follows:

$$|u_i(k)| \le I_{D_{max}} \tag{6.20}$$

where $I_{D_{max}}$ is the maximum allowed equalizing current for the DICM operated converter. In practice, we increase the duty cycle from 0 to 1 in 0.01 steps until we achieve the maximum allowed equalizing current that the converter operates in DICM. The allowable equalizing current for a DICM operated converter is linked to the converter's parameters and the PWM signal applied to it. By properly selecting the converter parameters or lowering the switching frequency of the PWM signal, the maximum allowed equalizing current in DICM can be increased [5].

Cell's current constraints: The i-th ($1 \le i \le n$) cell's current should be kept within a reasonable range to avoid detrimental excessive charging/discharging current, which can be expressed as:

$$|I_{B_i}(k)| \le I_{B_{max}} \tag{6.21}$$

where $I_{B_{max}}$ represents the maximum allowed cell's current. Based on (6.9) and (6.10), a suitable conservative treatment for the current constraints in (6.21) should satisfy the controlled equalizing current $u_i(k)$ in (6.17) as given below:

$$|u_i(k)| \le \tfrac{1}{2}(I_{B_{max}} - |I_s(k)|). \tag{6.22}$$

Combining (6.20) and (6.22), the controlled equalizing current of the i-th converter should be maintained in the limitation as follows:

$$|u_i(k)| \le I_{e_M}(k) \triangleq \min\{I_{D_{max}}, \tfrac{1}{2}(I_{B_{max}} - |I_s(k)|)\} \tag{6.23}$$

where $I_{e_M}(k)$ is defined as the maximum allowable cell equalizing current. According to (6.23), the bound of $u_i(k)$ varies with the change of the external current instead of remaining constant.

Remark 6.1 In the literature, the maximum permitted cell equalizing current is usually specified as a positive constant. The allowable cell equalizing current should be set as high as possible to reduce cell equalizing time. When the battery pack is charging/discharging with a large external current, the cell's current will easily exceed its bound in (6.21). This contradiction can be resolved by using (6.23), because the maximum allowed equalizing current varies with the external current.

6.3 SOC Estimation Based Quasi-Sliding Mode Control for Cell Equalization

6.3.1 Adaptive Quasi-Sliding Mode Observer Design for Cells' SOC Estimation

Because the SOC of the i-th ($1 \leq I \leq n$) cell in a serially connected battery pack is unmeasurable, an adaptive quasi-sliding mode observer is propounded to estimate it. To suppress chattering, the observer employs a continuous hyperbolic tangent function instead of the conventional discontinuous sign function, as shown in the following form:

$$\dot{\hat{x}}_i = A_i\hat{x}_i + b_i\varepsilon_i + e_i + h_i(y_i - c_i\hat{x}_i) + m_i\beta_i \tanh\left(\frac{y_i - c_i\hat{x}_i}{q}\right) \tag{6.24}$$

with the hyperbolic tangent function

$$\tanh\left(\frac{y_i - c_i\hat{x}_i}{q}\right) = \frac{e^{(y_i - c_i\hat{x}_i)/q} - e^{-(y_i - c_i\hat{x}_i)/q}}{e^{(y_i - c_i\hat{x}_i)/q} + e^{-(y_i - c_i\hat{x}_i)/q}}$$

where $\hat{x}_i \in \mathbb{R}^4$ denotes the estimation of x_i in (6.4), $h_i = [\, h_{i_1}, h_{i_2}, h_{i_3}, h_{i_4}]^T$ is the designed gain vector, q is a constant that satisfies $q \geq 0$, β_i is used to estimate the upper bound of the uncertain term φ_{i_M} and it can be updated as follows:

$$\dot{\beta}_i = \theta_i |y_i - c_i\hat{x}_i| - g_i\beta_i \tag{6.25}$$

where θ_i and g_i are designed positive constants. The minus term in (6.25) effectively increases the robustness of the update law [7]. Because it has a continuous form by using the smooth tanh(cdot) instead of the discontinuous sign function in the traditional sliding mode observer, the designed observer (6.24) is called a quasi-sliding mode observer [8] with the sliding surface converging in the vicinity of the origin instead of zero. From (6.4) and (6.24), the estimation error vector $\tilde{x}_i \triangleq x_i - \hat{x}_i$ can be derived as follows:

$$\dot{\tilde{x}}_i = (A_i - h_i c_i)\tilde{x}_i + m_i \varphi_i(x_i, \varepsilon_i) - m_i \beta_i \tanh(\frac{c_i \tilde{x}_i}{q}). \tag{6.26}$$

From [9], it is true for any $q > 0$ such that

$$0 \leq |c_i \tilde{x}_i| - c_i \tilde{x}_i \tanh(\frac{c_i \tilde{x}_i}{q}) \leq \kappa q \tag{6.27}$$

where κ is a positive constant such that $\kappa \approx 0.2785$.

Convergence analysis: A Lyapunov function is constructed to demonstrate the convergence of the designed adaptive quasi-sliding mode observer as follows:

$$V_i = \tilde{x}_i^T \tilde{x}_i + \frac{1}{\theta_i}(\beta_i - \varphi_{i_M})^2. \tag{6.28}$$

Since $\tilde{x}_i^T m_i = c_i \tilde{x}_i$ from (6.4), based on (6.5) and (6.27), the derivative of the Lyapunov function in (6.28) along the error trajectory yields

$$\dot{V}_i \leq -\tilde{x}_i^T Q_i \tilde{x}_i + M_{i1} \tag{6.29}$$

with

$$Q_i = (h_i c_i - A_i) + (h_i c_i - A_i)^T, \ M_{i1} = \frac{(\varphi_{i_M} g_i + \theta_i \kappa q)^2}{2 g_i \theta_i}.$$

From (6.4), (6.24), (6.29), we can say that a_{i_1}, a_{i_2}, and a_{i_3} are positive, Q_i is positive definite if

$$2a_{i_3} + 2h_{i_4} - \frac{n_{i_1}^2}{2a_{i_1}} - \frac{n_{i_2}^2}{2a_{i_2}} - \frac{h_{i_3}^2}{2a_{i_3}} \geq 0 \tag{6.30}$$

with

$$n_{i_1} = k_{2_i}(a_{i_3} - a_{i_1}) - h_{i_1}, n_{i_2} = a_{i_2} - a_{i_3} - h_{i_2}$$

holds. By making h_{i_4} enough large such that (6.30) is satisfied and Q_i is guaranteed to be positive definite. Because $\dot{V}_i$ is negative from outside the set $\{\|\tilde{x}_i\| \leq \sqrt{M_{i1}/\lambda_{min}(Q_i}\}$ with $\lambda_{min}(Q_i)$ is having the minimum eigenvalue of Q_i, V_i which decreases monotonically until the solutions enter the set $\{\|\tilde{x}_i\| \leq \sqrt{M_{i1}/\lambda_{min}(Q_i)}\}$ and cannot leave this set. In reference to the analysis of the boundedness in [10], the state estimation error of the designed adaptive quasi-sliding mode observer is proved to be bounded with the ultimate bound $\|\tilde{x}_i\| \leq \sqrt{M_{i1}/\lambda_{min}(Q_i)}$. Setting small q and g_i allows M_{i1} to be very small, resulting in a small sufficiently ultimate bound of the state estimation error to meet the requirement in practical applications.

Remark 6.2 Numerous SOC estimation algorithms, such as the sliding mode observers in [2], the multi-model extended Kalman filter in [11], and the nonlinear observer in [12], have been proposed with high accuracy in the literature.

The designed adaptive quasi-sliding mode observer has the following advantages over them. To begin with, the continuous term tanh(cdot) in (6.24) can retain the disturbance-suppression capacity of the non-smoothed sign function while suppressing unfavorable chattering phenomena. Besides which, the designed update law (6.25) can estimate an upper bound for the uncertain term without requiring its prior knowledge, increasing the disturbance-rejection capacity while yielding less conservation. Furthermore, as negative feedback, the minus term in (6.25) can significantly improve the robustness of the update law by ensuring that the estimation $beta_i$ remains bounded [7].

6.3.2 *Quasi-Sliding Mode-Based Cell Equalizing Control*

In this section, a discrete-time quasi-sliding mode control algorithm with the cell equalizing constraints described in the previous section is designed to have the modified Cûk converters collaborate efficiently to accomplish the cells' SOC equalization, based on the designed adaptive quasi-sliding mode observers.

Due to the inability to obtain the actual SOCs of the cells in the serially connected battery pack, the estimated SOCs of the designed adaptive quasi-sliding mode observers in (6.24) are used in the proposed cell equalizing algorithm. Based on the convergence analysis, the designed adaptive quasi-sliding mode observers can accurately estimate the SOCs of the cells, and the SOC estimation errors can converge to a small bound regardless of the initial SOC estimation errors. Our cell equalizing strategy is summarized below in order to eliminate the effect of initial SOC estimation errors. Prior to starting the cell equalizing controllers, we simply use the designed observers to estimate the SOCs of the cells. The cell equalizing controllers are turned on after a sufficient amount of time τ_0 that the cells' SOC estimation errors become small enough. As a result, during the cell equalizing process, the cells' SOC estimation errors can be assumed to be within a narrow range given as follows:

$$|\tilde{z}_i(k)| \leq \sigma, \forall k \geq \tau_0 \tag{6.31}$$

for all $1 \leq i \leq n$, where $\tilde{z}_i(k) = z_i(k) - \tilde{z}_i(k)$ with $z_i(k)$ and $\tilde{z}_i(k)$ are the i-th cell's actual and estimated SOCs, σ is a small positive constant, τ_0 is the instantaneous time at which the cell equalizing controllers started operating. With the initial SOC estimation errors unknown in practice, the delay τ_0 is required to guarantee the estimated SOCs in the designed cell equalizing algorithm within a limited range around the actual ones. A discrete-time quasi-sliding mode-based algorithm with saturated equalizing current constraints is designed as following to ensure the equalizing currents remain within the constraints in (6.23):

$$u(k) = -\ \mathrm{sat}(v(k)) - l(k)\eta\mathrm{sgn}(\alpha(k)) \tag{6.32}$$

with

$$l(k) = I_{e_M}(k), v(k) = \xi l(k) C\hat{z}(k),$$

$$\alpha(k) = \bar{C}^T C^T (C\hat{z}(k) - dC\bar{C}\text{sat}(v(k))),$$

$$C = \begin{bmatrix} -1 & 1 & 0 & \cdots & 0 \\ 0 & -1 & 1 & \cdots & 0 \\ 0 & 0 & -1 & \cdots & 0 \\ \cdots & \cdots & \cdots & \cdots & \cdots \\ 0 & 0 & 0 & \cdots & 1 \end{bmatrix}_{(n-1)\times n}$$

where $\hat{z}(k) = [\ \hat{z}_1(k), \ldots, \hat{z}_n(k)]^T \in \mathbb{R}^n$ represents the estimated SOC vector; $\xi > 1$ and η are the designed proportional gain; $I_{e_M}(k)$ is the maximum allowed cell equalizing current and it can be found in (6.23); $\text{sgn}(\alpha(k)) = [\ \text{sgn}(\alpha_1(k)), \ldots,$ $\text{sgn}(\alpha_{n-1}(k))]^T \in \mathbb{R}^{n-1}$ with $\text{sgn}(\alpha_i(k))$ is the sign function of $\alpha_i(k)$; $\text{sat}(v(k)) =$ $[\text{sat}(v_1(k)), \cdots, \text{sat}(v_{n-1}(k))]^T \in \mathbb{R}^{n-1}$, which satisfies

$$\text{sat}(v_i(k)) = \begin{cases} (1-\eta)l(k), & v_i(k) > (1-\eta)l(k) \\ v_i(k), & |v_i(k)| \le (1-\eta)l(k) \\ -(1-\eta)l(k), & v_i(k) < -(1-\eta)l(k). \end{cases}$$

To treat the saturation constraints more conveniently, the saturation function in (6.32) can be written as $\text{sat}(v(k)) = \delta(k)v(k)$ [13], with $\delta(k) = \text{diag}[\delta_1(k), \ldots, \delta_{n-1}(k)]$ and

$$\delta_i(k) = \begin{cases} (1-\eta)l(k)/v_i, & v_i(k) > (1-\eta)l(k) \\ 1, & |v_i(k)| \le (1-\eta)l(k) \\ -(1-\eta)l(k)/v_i, & v_i(k) < -(1-\eta)l(k). \end{cases} \tag{6.33}$$

From (6.33), it can be obtained that $0 < \delta_i(k) \le 1$ $(1 \le i \le n-1)$, which is represented as follows:

$$\rho I \le \delta(k) \le I \tag{6.34}$$

where I represents the identity matrix with suitable dimensions and ρ is a small positive constant. From (6.32), the sign of $u_i(k)$ in (6.32) is consistent with that of $\hat{z}_i(k) - \hat{z}_{i+1}(k)$ when $|\hat{z}_i(k) - \hat{z}_{i+1}(k)| \ge \eta/\xi$. The controlled cell equalizing current is selected as $u_i(k) = 0$, when the difference between the i-th and $(i+1)$-th cells' estimated SOCs converges to the range $(-\eta/\xi, \eta/\xi)$. The matrix $\bar{C}$ can then be obtained in advance for the controlled equalizing current design algorithm (6.32) based on the sign of the SOC difference between adjacent cells using (6.16). Generally, η/x_i is selected to be much smaller than the SOC estimation errors. Furthermore, when the differences between the cells' estimated SOCs and the average estimated SOC all converge to a tolerant range $(-\epsilon_1, \epsilon_1)$, we terminate the cell equalizing operation by setting $u(k) = 0_{n-1}$ with 0_{n-1} the column vector of $n-1$ zeros. This effectively avoids excessive equalization while also lowering the energy supply cost for cell equalizing circuits.

Theorem 6.1 *The serially connected cells' SOC differences can reach uniformly bounded under the algorithm (6.32) with $l(k)\eta 1_{n-1} \geq w(k)$, where 1_{n-1} is the column vector of $n-1$ ones.*

Proof A manifold is selected as $s(k) = Cz(k)$. We define $\tilde{s}(k) = C\tilde{z}(k)$ with $\tilde{z}(k) = z(k) - \hat{z}(k)$, from the cell equalizing model (6.17) and the input (6.32), it can be obtained that

$$\begin{aligned} s(k+1) = &(I - d\xi l(k)C\bar{C}\delta(k))s(k) + d\xi l(k)C\bar{C}\delta(k)\tilde{s}(k) \\ &- d\eta l(k)C\bar{C}\text{sgn}(\alpha(k)) + dC\bar{C}w(k) \end{aligned} \tag{6.35}$$

with $Cb(k) = 0_{n-1}$. Motivated by [14], a Lyapunov candidate is selected as $V_{22}(z(k)) = s^T(k)s(k)$. The change of the Lyapunov function is

$$\begin{aligned} \Delta V_{22} = &V_{22}(z(k+1)) - V_{22}(z(k)) \\ = &s^T(k)Qs(k) + 2d\xi l(k)s^T(k)N\delta(k)\tilde{s}^T(k) \\ &+ 2ds^T(k)Nw(k) - 2dl(k)\eta s^T(k)N\text{sgn}(\alpha(k)) \\ &+ 2d^2\xi l(k)\tilde{s}^T(k)\delta(k)Mw(k) + d^2w^T(k)Mw(k) \\ &- 2d^2l^2(k)\eta\xi s^T(k)\delta(k)M\text{sgn}(\alpha(k)) \\ &- 2d^2l(k)\eta\text{sgn}(\alpha^T(k))Mw(k) \\ &+ d^2l^2(k)\eta^2\text{sgn}^T(\alpha(k))M\text{sgn}(\alpha(k)) \\ &+ d^2\xi^2l^2(k)\tilde{s}^T(k)\delta(k)M\delta(k)\tilde{s}(k) \end{aligned} \tag{6.36}$$

with

$$M = \bar{C}^T C^T C\bar{C}, N = (I - d\xi l(k)\delta(k)\bar{C}^T C^T)C\bar{C},$$

$$Q = -d\xi l(k)C\bar{C}\delta(k) - d\xi l(k)\delta(k)\bar{C}^T C^T + d^2\xi^2 l^2(k)\delta(k)M\delta(k).$$

From (6.32) and (6.36), it yields $\alpha^T(k) = \hat{z}^T C^T N$. Since $w(k) \leq l(k)\eta 1_{n-1}$, it satisfies

$$-2d\eta l(k)\alpha^T(k)\text{sgn}(\alpha(k)) + 2d\alpha^T(k)w(k) \leq 0. \tag{6.37}$$

As $|w_i(k)| \leq w_M$ from (6.18) and $|\tilde{z}_i(k)| \leq \sigma$ from (6.31), it can be deduced that $\|\tilde{s}(k)\| \leq 2\sqrt{n-1}\sigma$ and $\|w(k)\| \leq \sqrt{n-1}w_M$, where n is the number of the cells in the battery pack. Using the Cauchy-Schwarz inequality [15], based on (6.32)–(6.37), it yields that

$$\Delta V_{22} \leq -s^T(k)Q_{22}s(k) + M_2 \tag{6.38}$$

with

$$Q_{22} = d\xi l(k)\delta(k)\bar{C}^T C^T - d^2\xi^2 l^2(k)\delta(k)M\delta(k),$$

$$M_2 = (n-1)(d^2(\eta^2 l^2(k) + w_M{}^2 + 2l(k)\eta w_M) \|M\|$$

$$+4\varpi\xi l(k)d\sigma^2 \|\delta(k)\| + 4\xi l(k)d^2\sigma(w_M + \eta l(k)) \|M\delta(k)\|$$

$$+4\xi^2 l^2(k)d^2\sigma^2 \|\delta(k)M\delta(k)\| + 4d\sigma(\eta l(k) + w_M) \|N\|)$$

where a suitable positive ϖ is chosen to make $d\xi l(k)s^T(k) \times C\bar{C}\delta(k)(1/\varpi\bar{C}^T C^T - I)s(k) \leq 0$. The maximum limit of the cell equalizing current $l(k)$ is selected as much less than $1/d$, since d has a very small value that satisfies $d = \eta_0 T/C_b$ with $C_b = 3600 \times$ Ampere-hour capacity. Since, $C\bar{C}$ is positive definite and the expression with term d^2 is much less than that with d, Q_{22} in (6.38) is positive definite. Furthermore, based on (6.38) and the boundedness analysis in [10], the cells' SOC differences are uniformly bounded with the ultimate bound $\|s\| \leq \sqrt{M_2/\lambda_{min}(Q_{22})}$, where $\lambda_{min}(Q_{22})$ is the minimum eigenvalue of Q_{22}. As the SOC estimation error σ bound is small, the terms including $d\sigma$ and d^2 in M_2 are much smaller than those with d in Q_{22}. Then, it can be found that M_2 is small enough to be compared with $\lambda_{min}(Q_{22})$. As a result, under the designed SOC estimation-based quasi-sliding mode control with saturated equalizing current constraints, the cells' actual SOC differences can converge to a small bound, satisfying the performance demand in practice. The following experiments will validate the bound of the cells' SOC differences.

6.4 Experiments

6.4.1 Experimental Setup

In the experiment, a four-modular serially connected lithium-ion battery pack with three bidirectional modified Cûk converters is used. Experiment with the sampling and control signals generated by an NI GPIC Single-Board 9683 on the test bench is shown in Fig. 6.3a.

Battery cells In Fig. 6.3b, four NCR 18650 (MH12210) lithium-ion batteries are illustrated. After conducting many charging and discharging experiments, the capacities of those cells are identified as 3.1 Ah ($C_b = 3.1 \times 3600$ F) and the Coulombic efficiency η_c is 0.96. A similar test procedure is validated in [1] for extracting the equivalent circuit parameters in the i-th ($1 \leq i \leq 4$) battery model (6.4). The OCVs and the SOCs relation between the cells are depicted in Fig. 6.4a. The average model parameters of the four cells are used with $k_{1_i} = 3.231$, $k_{2_i} = 0.8948$, $R_{0_i} = 0.206\ \Omega$, $R_{s_i} = 0.0158\ \Omega$, $C_{s_i} = 12304$ F, $R_{f_i} = 0.0151\ \Omega$, and $C_{f_i} = 1584$ F ($1 \leq i \leq 4$) for the designed adaptive quasi-sliding mode based SOC observers. The unknown model disturbances are the differences between the cell's model parameters and the average

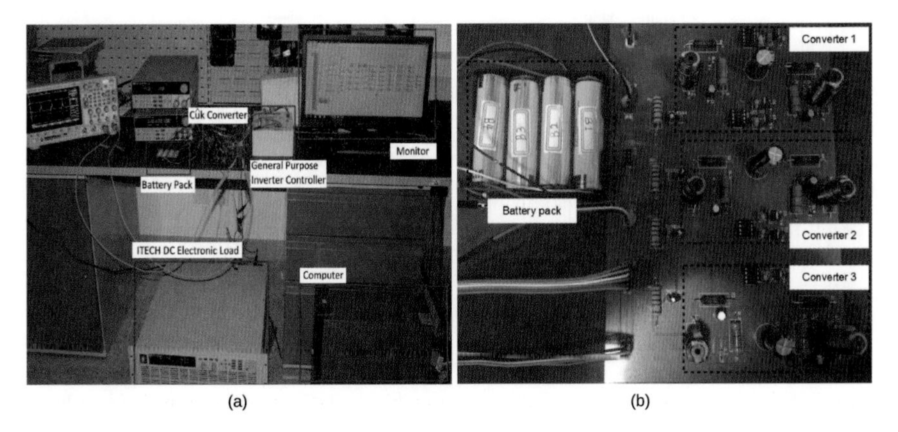

Fig. 6.3 **a** Experimental test bench, **b** battery cells and bidirectional modified Cûk converter board

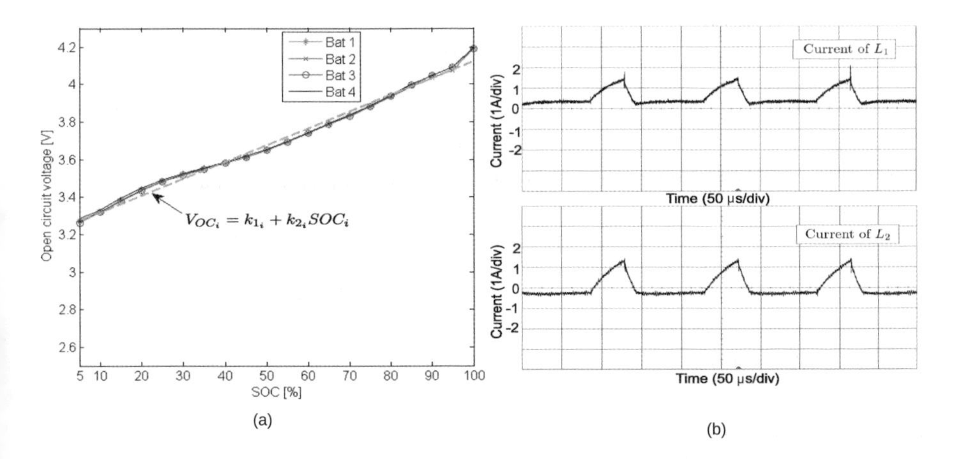

Fig. 6.4 **a** Relationship between the open circuit voltage and the SOC, **b** inductor currents of the modified Cûk converter

parameters. For more accurate SOC estimation, we can calibrate the cells' capacities and model parameters on a regular basis.

A series of prior simulations in Simulink/MATLAB are performed based on a battery with the model parameters extracted above to find suitable parameters of the designed adaptive quasi-sliding mode observer for the trade-off between chattering and convergence speed. A FUDS [16] is utilized for the battery experiment with the current profile depicted in Fig. 6.5a. The actual and estimated SOCs initial values are selected as 70% and 55%, respectively. The comparison of SOC estimation errors based on the different selection of parameters of the designed adaptive quasi-sliding mode observer is shown in Fig. 6.5b. It illustrated that increasing the observer's convergence speed causes the chattering to become more intense. The parameters

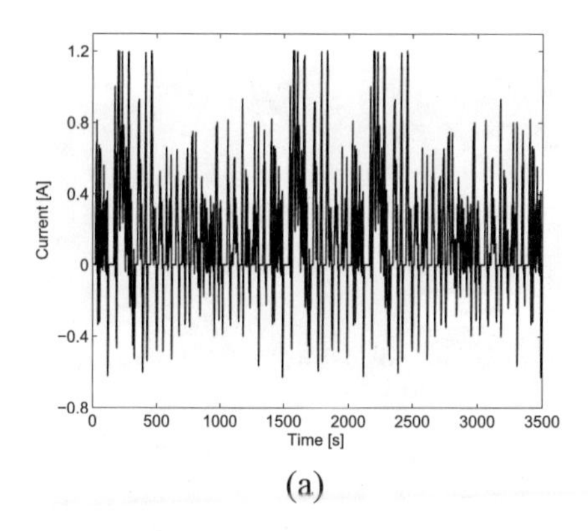

(a)

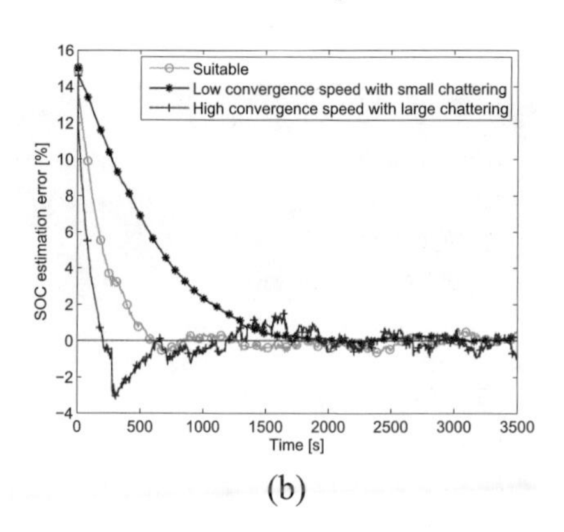

(b)

Fig. 6.5 Simulation results of **a** the cell's current profile, **b** SOC estimation errors with different selected parameters for the designed observer

$h_i = [\,0.15, 0.15, 0.15, 1.5]^T$ and $q = 0.02$ are selected for the observers to balance convergence speed and accuracy, with simulation performance shown as the indigo line in Fig. 6.5b. Higher sampling rates can result in more precise signal acquisition but increase the operating burden of the controller, so the sampling period of the observer is set to 0.1 s as a trade-off in the experiments. The simulation results provide the roughly consumed time range in which the cell's SOC estimation error enters its convergence bound, and a suitable turn-on time of the cell equalizing controllers $\tau_0 = 600$ s that is greater than the upper bound of the consumed SOC estimation time is selected in the following experiments.

Bidirectional modified Cûk converters: Three self-developed bidirectional modified Cûk converters are shown in Fig. 6.3b. The i-th ($1 \leq i \leq 3$) converter's parameters are chosen as $L_{i1} = L_{i2} = 100\ \mu$H and $C_i = 470\ \mu$F [5]. NTD6416AN-1G MOSFETs are driven by the PWM signals with a 7 kHz switching frequency. A preliminary experiment is carried out to test the performance of a bidirectional modified Cûk converter that connects two cells. The terminal voltages of its connected cells' are $V_{B_1} = 3.93$ V and $V_{B_2} = 3.62$ V, respectively. The PWM signal applied to MOSFET Q_1 has a duty cycle of 0.3. In Fig. 6.4b, the transient inductor current curves recorded by a Keysight oscilloscope are denoted by measuring the voltages of the resistors serially connected with the inductors. With an energy transfer efficiency of 0.878, the average inductor currents are $I_{L_1} = 0.147$ A and $I_{L_2} = 0.129$ A. It is consistent with the converter analysis.

The cell's maximum allowed current is set as $I_{B_{max}} = 3$ A, and the maximum allowed equalizing current for the converter operated in DICM is chosen as $I_{D_{max}} = 1$ A. The sampling period is $T = 2$ s. The termination is selected as $\epsilon_1 = 1\%$. For the proposed discrete-time quasi-sliding mode based cell equalizing algorithm, the gain η is set to 0.02 and ξ is chosen as 15, respectively. Note that other suitable values can also be assigned, depending on the needs and performance in practice.

6.4.2 Experimental Results

Cell equalizing with a standby mode: Experiments for the cell equalizing issue are validated while the battery pack is in standby mode. The cells' initial SOCs in the battery pack are $SOC_1(0) = 74\%$, $SOC_2(0) = 82\%$, $SOC_3(0) = 71\%$, and $SOC_4(0) = 80\%$, which can be obtained by discharging them from the fully charged states. All of the cells' initial estimated SOCs for the designed adaptive quasi-sliding mode observers are set to 75%. The cells' actual and estimated SOCs are shown in Fig. 6.6a, where the cells' actual SOCs are achieved by using the ampere-hour counting system with actual initial SOCs [2]. The SOC estimation errors are all less than $\pm 0.5\%$ after 600 s, as shown in Fig. 6.6a. After about 2286 s (including the 600 s before the controllers are started), the differences between the cells' actual SOCs and the average actual SOC can converge to the bound $(-1.07\%, 0.88\%)$, which can satisfy the needs in practice. The corresponding PWM duty cycles are shown in Fig. 6.7a. The maximum allowed cell equalizing current is a constant because the external current is zero.

Cell equalizing with a discharging mode: The battery pack is connected with an ITECH programmable DC electronic load with the discharging current illustrated in Fig. 6.8. The cells' initial SOCs are set as $SOC_1(0) = 84\%$, $SOC_2(0) = 92\%$, $SOC_3(0) = 81\%$, and $SOC_4(0) = 88\%$, respectively. The initial estimated cells' SOCs of the designed observers are all set as 75%.

The results of the designed SOC estimation based quasi-sliding mode control algorithm are depicted in Fig. 6.9a. The SOC estimation errors converge to a bound less than $\pm 1\%$, indicating that the designed adaptive quasi-sliding mode observers perform exceptionally well.

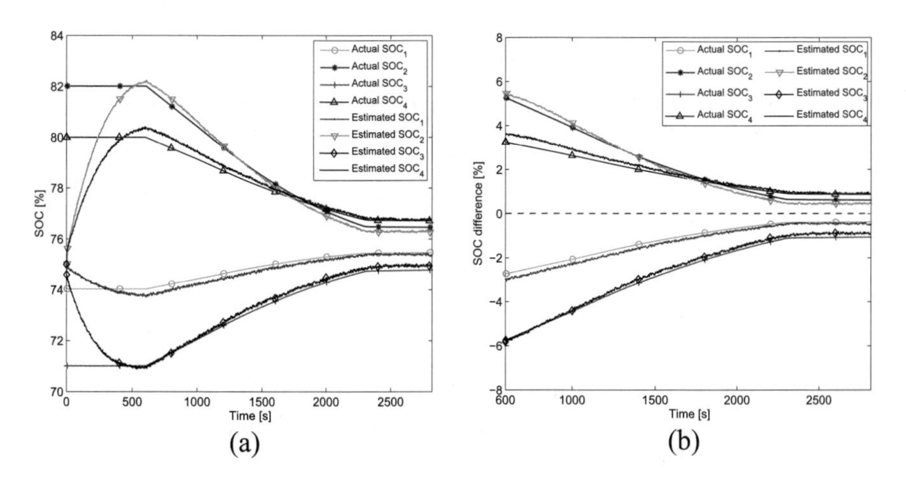

Fig. 6.6 **a** Cells' actual and estimated SOCs, **b** differences between the cells' SOCs and the average actual SOC under the proposed SOC estimation based cell equalizing algorithm when the battery pack is in a standby mode

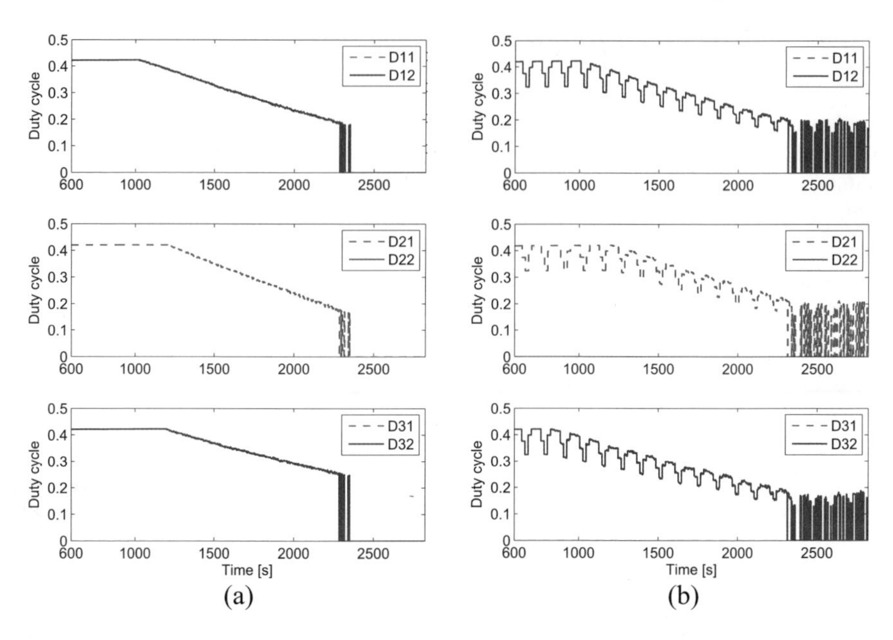

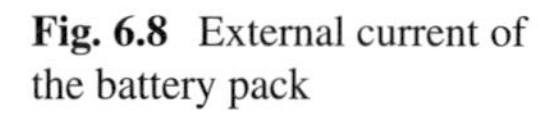

Fig. 6.7 PWM duty cycles under the proposed SOC estimation based cell equalizing algorithm **a** when the battery pack is in a standby mode, **b** when the battery pack is in a discharging mode

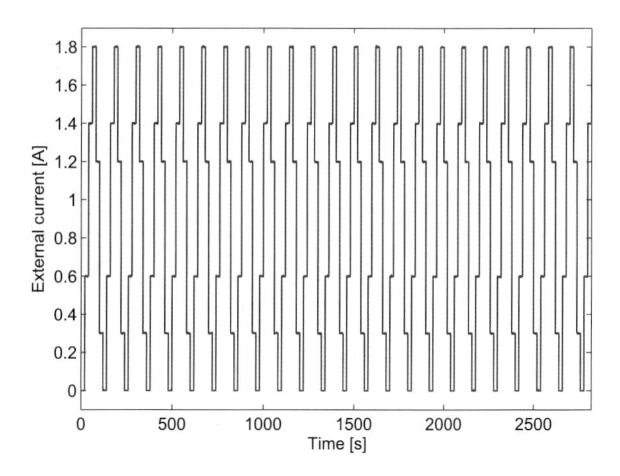

Fig. 6.8 External current of the battery pack

As shown in Fig. 6.9b, the differences between the cells' actual SOCs and the average actual SOC can converge to the bound (−0.9%, 1.1%) after about 2340 s (such as the consumed SOC estimation period prior to the cell equalizing process). The corresponding PWM duty cycles are illustrated in Fig. 6.7b, that vary with the transformation in external current to protect the cells from drawing too much current. As shown in Figs. 6.7b and 6.9b, SOC estimation errors cause more disturbance for the propounded SOC estimation-based quasi-sliding mode control algorithm as the

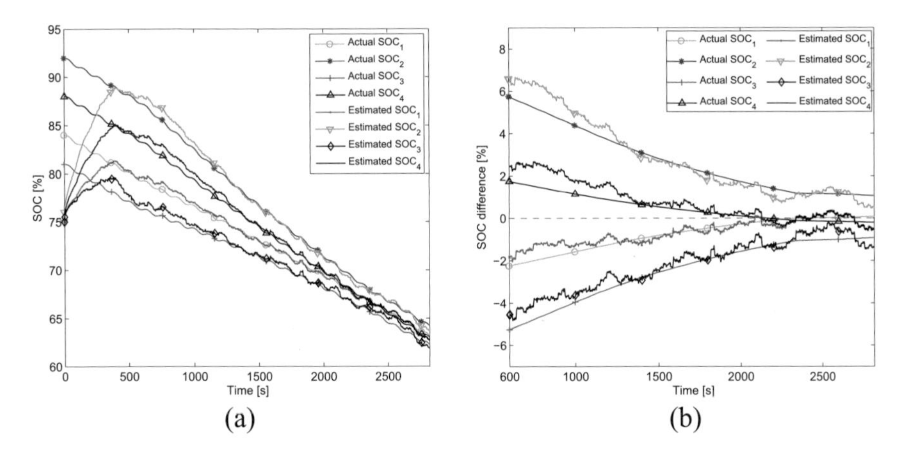

Fig. 6.9 **a** Cells' actual and estimated SOCs, **b** differences between the cells' SOCs and the average actual SOC under the designed SOC estimation based cell equalizing algorithm when the battery pack is in a discharging mode

actual SOCs of adjacent cells decrease. Excessive cell equalization can be avoided and the energy supply cost for cell equalizers reduced by terminating the cell equalizing process when the cells' estimated SOC differences and the average estimated SOC are all within the tolerant range $(-1\%, 1\%)$. Because the absolute values of the SOC estimation errors are less than 1%, it is reasonable to set the termination $\epsilon_1 = 1\%$. The experimental results show that the proposed SOC estimation-based quasi-sliding mode control algorithm performs magnificently for serially connected battery pack cell equalization.

References

1. M. Chen, G. Rincon-Mora, Accurate electrical battery model capable of predicting runtime and I–V performance. IEEE Trans. Energy Convers. **21**(2), 504–511 (2006)
2. M. Gholizadeh, F.R. Salmasi, Estimation of state of charge, unknown nonlinearities, and state of health of a lithium-ion battery based on a comprehensive unobservable model. IEEE Trans. Ind. Electron. **61**(3), 1335–1344 (2013)
3. G.L. Plett, Extended Kalman filtering for battery management systems of LiPB-based HEV battery packs: Part 2: modeling and identification. J. Power Sources **134**(2), 262–276 (2004)
4. Y.-S. Lee, M.-W. Cheng, Intelligent control battery equalization for series connected lithium-ion battery strings. IEEE Trans. Ind. Electron. **52**(5), 1297–1307 (2005)
5. L. Yuang-Shung, D. Jiun-Yi, Fuzzy-controlled individual-cell equaliser using discontinuous inductor current-mode Cuk convertor for lithium-ion chemistries. IEE Proc.-Electr. Power Appl. **152**(5), 1271–1282 (2005)
6. Q. Ouyang, J. Chen, C. Xu, H. Su, Cell balancing control for serially connected Lithium-ion batteries, in *Proceedings American Control Conference*, (Boston, MA, USA, 2016), pp. 3095–3100

7. Y. Hong, H. Qin, G. Chen, Adaptive synchronization of chaotic systems via state or output feedback control. Int. J. Bifurc. Chaos **11**(4), 1149–1158 (2001)
8. Y. Shtessel, C. Edwards, L. Fridman, A. Levant, *Sliding Mode Control and Observation* (Birkhäuser New York, NY, Springer, 2014)
9. M.M. Polycarpou, P.A. Ioannou, A robust adaptive nonlinear control design, in *American Control Conference*, vol. 1993 (San Francisco, CA, USA,1993), pp. 1365–1369
10. H.K. Khalil, *Nonlinear Systems* (Prentice Hall, Englewood Cliffs, NJ, USA, 2002)
11. H. Fang, X. Zhao, Y. Wang, Z. Sahinoglu, T. Wada, S. Hara, R.A. de Callafon, Improved adaptive state-of-charge estimation for batteries using a multi-model approach. J. Power Sources **254**, 258–267 (2014)
12. J. Chen, Q. Ouyang, C. Xu, H. Su, Neural network-based state of charge observer design for lithium-ion batteries. IEEE Trans. Control Syst. Technol. **26**(1), 313–320 (2017)
13. Z. Zhu, Y. Xia, M. Fu, Adaptive sliding mode control for attitude stabilization with actuator saturation. IEEE Trans. Ind. Electron. **58**(10), 4898–4907 (2011)
14. H. Zhang, F.L. Lewis, Z. Qu, Lyapunov, adaptive, and optimal design techniques for cooperative systems on directed communication graphs. IEEE Trans. Ind. Electron. **59**(7), 3026–3041 (2012)
15. M. Masjed-Jamei, A functional generalization of the Cauchy-Schwarz inequality and some subclasses. Appl. Math. Lett. **22**(9), 1335–1339 (2009)
16. H. He, R. Xiong, X. Zhang, F. Sun, J. Fan, State-of-charge estimation of the lithium-ion battery using an adaptive extended Kalman filter based on an improved Thevenin model. IEEE Trans. Veh. Technol. **60**(4), 1461–1469 (2011)

Chapter 7
Module-Based Cell-to-Cell Equalization Control

7.1 Module-Based Cell-to-Cell Equalization Systems

A module-based cell-to-cell equalizing method [1, 2] is used in this instance, as shown in Fig. 7.1, where a battery pack is split into m battery modules, each of which has n cells. Energy balancing between battery modules is accomplished by m-1 module-level equalizers, while energy equalization inside each battery module is accomplished by n-1 cell-level equalizers. It should be noted that real-world applications have successfully used this kind of equalization circuitry [3]. One of these often employed equalizers is the modified bidirectional Cûk converter, which is used in our study as seen in Fig. 7.2. Its operation is clear from [4, 5]. It should be noted that the suggested technique may be easily modified to be used to various types of cell-to-cell equalizers, with the major focus of this study being on the design of the appropriate cell equalizing current control algorithm.

7.1.1 Equalizing Currents

As shown in Fig. 7.2, the j-th ($1 \leq j \leq n-1$) cell-level equalizer in the i-th ($1 \leq i \leq m$) battery module may transmit energy in both directions between its left and right adjacent cells, j and j+1, using the equalizing currents $I_{cel_{i,j}}(k)$ and $I_{cer_{i,j}}(k)$. By controlling the equalizing currents through their associated equalizer, the energy should be transported from the cells with higher SOC to the cells with lower SOC in order to accomplish cell equalization. By comparing the SOC of the j-th and (j+1)-th cells in the i-th battery module, it is possible to predict the direction of the equalizing current through the j-th equalizer. In this case, $I_{cel_{i,j}}(k)/I_{cer_{i,j}}(k)$ is chosen as the controlled equalizing current if the SOC of the j-th/ (j+1)-th cell is higher than

J. Chen et al., *Equalization Control for Lithium-Ion Batteries*,
https://doi.org/10.1007/978-981-99-0220-0_7

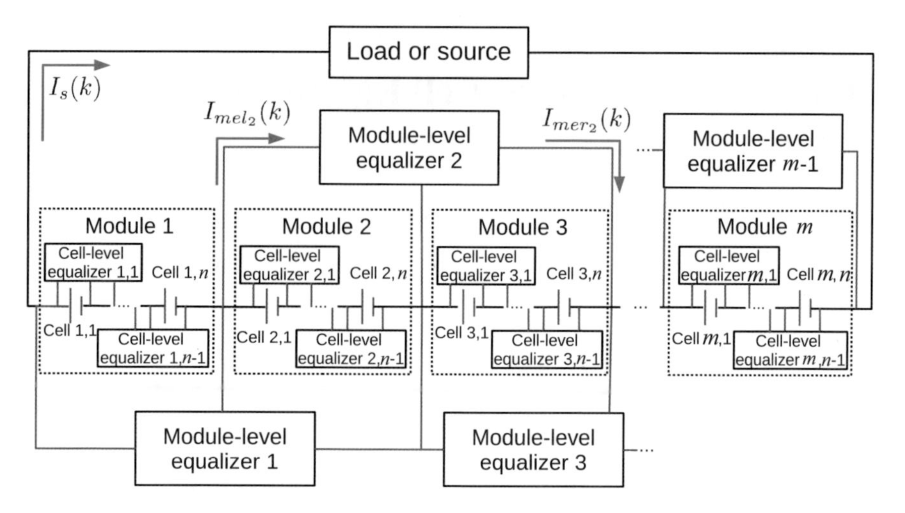

Fig. 7.1 The module-based cell-to-cell equalizing system for a battery pack of mn serially linked cells is shown in the diagram

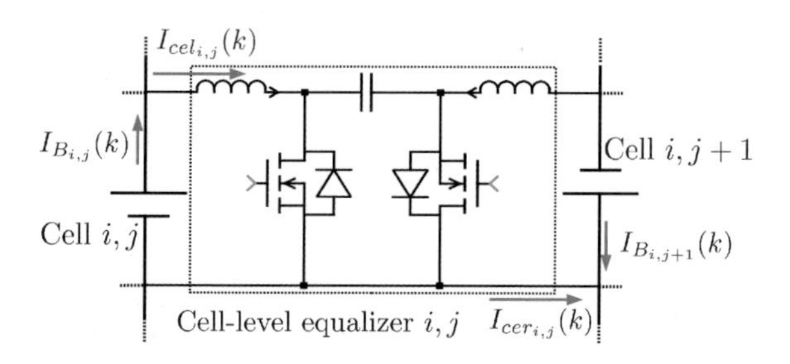

Fig. 7.2 Cell equalizer based on modified bidirectional Cûk converter

the SOC of the $(j+1)$-th/j-th; $I_{cel_{i,j}}(k)$ and $I_{cer_{i,j}}(k)$ can therefore be replaced by a new non-negative variable, $I_{ce_{i,j}}(k)$ as:

$$
\begin{aligned}
I_{cel_{i,j}}(k) &= (k_{i,j} + k'_{i,j}\beta_{i,j})I_{ce_{i,j}}(k) \\
I_{cer_{i,j}}(k) &= (k_{i,j}\beta_{i,j} + k'_{i,j})I_{ce_{i,j}}(k)
\end{aligned}
\tag{7.1}
$$

with

$$
\begin{cases}
k_{i,j} = 1, k'_{i,j} = 0, & \text{if } SOC_{i,j}(k) \geq SOC_{i,j+1}(k) \\
k_{i,j} = 0, k'_{i,j} = -1, & \text{if } SOC_{i,j}(k) < SOC_{i,j+1}(k)
\end{cases}
$$

where $\beta_{i,j}$ $(0 < \beta_{i,j} \leq 1)$ is the ratio of the equalizing currents $I_{cel_{i,j}}(k)$ and $I_{cer_{i,j}}(k)$, which is known as the equalizer's energy transfer efficiency [4, 6], and $I_{ce_{i,j}}(k)$ denotes the controlled equalizing current of the j-th cell-level equalizer in the i-th battery module. $SOC_{i,j}(k)$ is the SOC of the j-th cell in the i-th battery module. Keep in mind that the equalizing current through the j-th cell-level equalizer is defined by the current direction out/into of the j-th/ (j+1)-th cell (as the arrow direction in Fig. 7.2), and the negative sign in $k'_{i,j}$ in (7.1) indicates that its direction is opposite to the predefined reference direction.

Similar to this, energy may be moved from battery modules with higher SOC to modules with lower SOC using module-level equalizers; the SOC of a battery module is determined by the average SOCs of its cells. The left and right neighboring i-th and (i+1)-th battery modules of the i-th ($1 \leq i \leq m-1$) module-level equalizer's equalizing currents can be expressed as follows:

$$\begin{aligned} I_{mel_i}(k) &= (k_{m_i} + k'_{m_i}\beta_{m_i}) I_{me_i}(k) \\ I_{mer_i}(k) &= (k_{m_i}\beta_{m_i} + k'_{m_i}) I_{me_i}(k) \end{aligned} \tag{7.2}$$

with

$$\begin{cases} k_{m_i} = 1, k'_{m_i} = 0, & \text{if } SOC_{m_i}(k) \geq SOC_{m_{i+1}}(k) \\ k_{m_i} = 0, k'_{m_i} = -1, & \text{if } SOC_{m_i}(k) < SOC_{m_{i+1}}(k) \end{cases}$$

where $I_{me_i}(k)$ and β_{m_i} stand for the controlled equalizing current and energy transfer efficiency of the i-th module-level equalizer, respectively; $SOC_{m_i}(k) = \frac{1}{n}\sum_{j=1}^{n} SOC_{i,j}(k)$ is the i-th battery module's SOC; $I_{mel_i}(k)$ and $I_{mer_i}(k)$ are the i-th and (i+1)-th battery modules' equalizing currents through the i-th module-level equalizer, with the reference directions being defined as the directions of the arrows in Fig. 7.1.Keep in mind that each battery module's cells all share the same module-level equalizing current because of the connection's serial nature. The equalizing current for the cell, as seen in Fig. 7.1, is made up of the currents flowing through the coupled module-level and cell-level equalizers and may be computed as follows:

$$\begin{aligned} I_{e_{1,1}}(k) &= I_{cel_{1,1}}(k) + I_{mel_1}(k) \\ I_{e_{1,j}}(k) &= -I_{cer_{1,j-1}}(k) + I_{cel_{1,j}}(k) + I_{mel_1}(k) \\ I_{e_{1,n}}(k) &= -I_{cer_{1,n-1}}(k) + I_{mel_1}(k) \\ I_{e_{i,j}}(k) &= -I_{cer_{i,j-1}}(k) + I_{cel_{i,j}}(k) - I_{mer_{i-1}}(k) + I_{mel_i}(k) \\ I_{e_{m,1}}(k) &= I_{cel_{m,1}}(k) - I_{mer_{m-1}}(k) \\ I_{e_{m,j}}(k) &= -I_{cer_{m,j-1}}(k) + I_{cel_{m,j}}(k) - I_{mer_{m-1}}(k) \\ I_{e_{m,n}}(k) &= -I_{cer_{m,n-1}}(k) - I_{mer_{m-1}}(k) \end{aligned} \tag{7.3}$$

for $2 \leq i \leq m-1, 2 \leq j \leq n-1$, where $I_{e_{i,j}}(k)$ is the equalizing current for the j-th cell in the i-th battery module, which is characterized as positive/negative when it discharges/charges the cell. Then, it may be rewritten as follows by substituting (7.1) and (7.2) into (7.3):

$$
\begin{aligned}
I_{e_{1,1}}(k) &= p_{1,1} I_{ce_{1,1}}(k) + p_{m_1} I_{me_1}(k) \\
I_{e_{1,j}}(k) &= -p'_{1,j-1} I_{ce_{1,j-1}}(k) + p_{1,j} I_{ce_{1,j}}(k) + p_{m_1} I_{me_1}(k) \\
I_{e_{1,n}}(k) &= -p'_{1,n-1} I_{ce_{1,n-1}}(k) + p_{m_1} I_{me_1}(k) \\
I_{e_{i,j}}(k) &= -p'_{i,j-1} I_{ce_{i,j-1}}(k) + p_{i,j} I_{ce_{i,j}}(k) + p_{m_i} I_{me_i}(k) - p'_{m_{i-1}} I_{me_{i-1}}(k) \\
I_{e_{m,1}}(k) &= p_{m,1} I_{ce_{m,1}}(k) - p'_{m_{m-1}} I_{me_{m-1}}(k) \\
I_{e_{m,j}}(k) &= -p'_{m,j-1} I_{ce_{m,j-1}}(k) + p_{m,j} I_{ce_{m,j}}(k) - p'_{m_{m-1}} I_{me_{m-1}}(k) \\
I_{e_{m,n}}(k) &= -p'_{m,n-1} I_{ce_{m,n-1}}(k) - p'_{m_{m-1}} I_{me_{m-1}}(k)
\end{aligned}
\tag{7.4}
$$

with $p_{i,j} = k_{i,j} + k'_{i,j}\beta_{i,j}$, $p'_{i,j} = k_{i,j}\beta_{i,j} + k'_{i,j}$, $p_{m_i} = k_{m_i} + k'_{m_i}\beta_{m_i}$, $p'_{m_i} = k_{m_i}\beta_{m_i} + k'_{m_i}$. It should be noted that the gains $p_{i,j}$, $p'_{i,j}$, p_{m_i}, and p'_{m_i} in (7.4) may be predetermined by comparing the current SOCs of neighboring battery cells and modules, and they can be used as known parameters in the construction of the next cell equalizing control algorithm.

7.1.2 *Cell Equalizing System Model*

The battery pack's SOC, which is computed as the difference between the cell's remaining capacity and its nominal capacity, may be calculated for the j-th ($1 \leq j \leq n$) cell in the i-th ($1 \leq i \leq m$) module as follows [7]:

$$
SOC_{i,j}(k+1) = SOC_{i,j}(k) - dI_{B_{i,j}}(k) \tag{7.5}
$$

with $d = \frac{\eta_0 T}{Q}$, where Q stands for the nominal capacity of the cells, which is typically chosen uniformly across the board for every cell in the battery pack [3, 6]; the sample period is T and the Coulombic efficiency is η_0, respectively, the current of the j-th cell in the i-th battery module is described as $I_{B_{i,j}}(k)$, which is positive or negative depending on whether the cell is in the discharging or charging phase. We can obtain the following:

$$
I_{B_{i,j}}(k) = I_s(k) + I_{e_{i,j}}(k) \tag{7.6}
$$

where $I_s(k)$ stands for the external current (provided by the charger or to the load) that the battery pack supplies. The SOCs of the cells may be well determined by applying the current SOC estimating methods described in literatures like [8–11]. The model of the cell equalizing system for the battery pack made up of mn serially linked cells may be expressed as the discrete state space form as given below which is based on (7.4–7.6).

$$
x(k+1) = x(k) + d(C_1 u_1(k) + C_2 u_2(k) - b(k)) \tag{7.7}
$$

where the state vector is $x(k) = [x_1(k), \cdots, x_m(k)] \in \mathbb{R}^{mn}$ with $x_i(k) \triangleq [SOC_{i,1}(k), \cdots, SOC_{i,n}(k)] \in \mathbb{R}^n$ denoting the cells' SOC vector in the i-th battery module; the input vectors $u_1(k) \triangleq [I_{me_1}(k), \cdots, I_{me_{m-1}}(k)]^T \in \mathbb{R}^{m-1}$ and $u_2(k) = [u_{2,1}(k), \cdots, u_{2,m}(k)]^T \in \mathbb{R}^{m(n-1)}$ with $u_{2,i}(k) \triangleq [I_{ce_{i,1}}(k), \cdots, I_{ce_{i,n-1}}(k)]^T \in \mathbb{R}^{n-1}$ are the controlled equalizing currents of the module-level and cell-level equalizers, respectively; $b(k) = [b_1(k), \cdots, b_m(k)]^T \in \mathbb{R}^{mn}$ with $b_i = [I_s(k), \cdots, I_s(k)]^T \in \mathbb{R}^n$; $C_1(k) \in \mathbb{R}^{mn\times(m-1)}$ and $C_2(k) \in \mathbb{R}^{mn\times m(n-1)}$ are

$$C_1 = \begin{bmatrix} -C_{1,1} & 0_n & \cdots & 0_n \\ C'_{1,1} & -C_{1,2} & \cdots & 0_n \\ \cdots & \cdots & \cdots & \cdots \\ 0_n & 0_n & \cdots & C'_{1,m-1} \end{bmatrix}$$

$$C_2 = \begin{bmatrix} C_{2,1} & 0_{n,n-1} & \cdots & 0_{n,n-1} \\ 0_{n,n-1} & C_{2,2} & \cdots & 0_{n,n-1} \\ \cdots & \cdots & \cdots & \cdots \\ 0_{n,n-1} & 0_{n,n-1} & \cdots & C_{2,m} \end{bmatrix}$$

where 0_n represents the column vector with n zeros and $0_{n,n-1}$ denotes the zero matrix with dimension of $n \times (n-1)$, respectively; $C_{1,i} \in \mathbb{R}^n$, $C'_{1,i} \in \mathbb{R}^n$, and $C_{2,i} \in \mathbb{R}^{n\times(n-1)}$ are defined as $C_{1,i} = [p_{m_i}, \cdots, p_{m_i}]^T$, $C'_{1,i} = [p'_{m_i}, \cdots, p'_{m_i}]^T$, and

$$C_{2,i} = \begin{bmatrix} -p_{i,1} & 0 & \cdots & 0 \\ p'_{i,1} & -p_{i,2} & \cdots & 0 \\ \cdots & \cdots & \cdots & \cdots \\ 0 & 0 & \cdots & p'_{i,n-1} \end{bmatrix}.$$

It should be noted that the model in (7.7) that can be used to a wide variety of cells and equalizers regardless of type is used to define the states of cells without taking into account each cell's internal responses. As a result, the cell equalizing control method developed using this model is more adaptable and practical.

7.1.3 Cell Equalizing Constraints

To maintain the safety and security of the cell equalizing system, the following hard constraints including the SOCs and currents of the cells and the equalizing currents that the equalizers can supply must be carefully taken into account.

SOC restrictions: To prevent overcharging and overdischarging, the cells' SOCs should be kept within a normal operating range with

$$x_l 1_{mn} \leq x(k) \leq x_u 1_{mn} \tag{7.8}$$

where x_l and x_u signify the lower and upper limits of the cells' SOCs, respectively.

Equalizing current limitations: The maximum permitted equalizing current that an equalization can provide is written as

$$\begin{aligned} 0_{m-1} &\le u_1(k) \le I_{me_u} 1_{m-1} \\ 0_{m(n-1)} &\le u_2(k) \le I_{ce_u} 1_{m(n-1)} \end{aligned} \tag{7.9}$$

where I_{me_u} and I_{ce_u} are the maximum permitted equalizing currents that the module-level and cell-level equalizers, respectively, may deliver.

Current constraints: The threshold of currents is critical to the safety of the cells since excessive current might impair their functionality or possibly ignite a fire. As a result, it is recommended that the cells' currents be kept within a range that satisfy the following:

$$I_{B_c} \le I_{B_{i,j}}(k) \le I_{B_d} \tag{7.10}$$

for $1 \le i \le m$ and $1 \le j \le n$, where I_{B_c} and I_{B_d} are the maximum permitted charging and discharging currents of the cells, respectively, and in which I_{B_d} is positive and I_{B_c} is negative. The regulated equalizing currents $u_1(k)$ and $u_2(k)$ in (7.7) should meet the condition in (7.10) by replacing the expression from (7.6) into (7.10).

$$I_{B_c} 1_{mn} + b(k) \le C_1 u_1(k) + C_2 u_2(k) \le I_{B_d} 1_{mn} + b(k). \tag{7.11}$$

It should be noted that in the literature, such as [5, 12–14], the effect of external current on the cell equalizing system is disregarded and the cells' maximum permitted equalizing current is specified as constant. When the battery pack is experiencing a significant external current, the cell currents might rapidly surpass their limit. This shortcoming may be successfully fixed by using (7.11), which changes the bound of the equalizing currents with the change of the external current. Furthermore, from (7.7), it notices that the module-level and cell-level equalizers attached to it have an impact on the SOCs of the cells. This shows the superiority of treating the cell equalizing system as a whole system here as opposed to the literature, which mainly focuses on individually controlling the equalizer for only cell equalizing between two neighboring cells. Larger equalizing currents with satisfying (7.11) can be obtained by cooperatively controlling these equalizers to shorten the equalizing time.

7.2 Hierarchical Optimal Control Strategy

In this section, the cell equalizing task is first formulated in light of the model and the rigid constraints of the module-based cell equalizing system from the previous section. This is followed by a thorough explanation of our suggested hierarchical cell equalizing control strategy for the battery pack.

7.2.1 Cell Equalizing Task Formulation

Cell equalization's primary goal is to quickly get all of the battery pack's cells' SOCs to the same level. It is intended to reduce the difference between the SOCs of the cells and their average value in order to achieve this cell equalizing objective. The corresponding cost function can be written as

$$J_e = (x(k) - \bar{x}(k))^T (x(k) - \bar{x}(k)) \tag{7.12}$$

where $\bar{x}(k) \in \mathbb{R}^{mn}$ is the average SOC vector of the cells in the battery pack, satisfying

$$\bar{x}(k) = \frac{1}{mn} 1_{mn} 1_{mn}^T x(k) \tag{7.13}$$

where 1_{mn} represents the column vector of mn ones.

To keep the cells safe, the temperature rise during the equalization process should be constrained. According to [15], the currents of the cells can be limited at each sampling step to prevent the temperature from rising too much. This is because a greater cell's current causes a higher temperature rise. Based on (7.6), the size of the regulated equalizing currents $u_1(k)$ and $u_2(k)$ can be restricted to lessen the temperature rise in the cells since the external current is not the control variable for cell equalizing. Consequently, a cost function that

$$J_l = u_1^T(k) u_1(k) + u_2^T(k) u_2(k). \tag{7.14}$$

Be aware that limiting the equalizing currents, like in the case of the cost function, can also prevent the hardware equalizer circuits' temperature from rising (7.14).

Multi-objective formulation: For the cell equalization problem, the suppression of temperature rise and the decrease of equalizing time are both taken into account. The multi-objective cost function may be created by fusing (7.12) and (7.14) as follows to balance them out:

$$J = \gamma_1 J_e + \gamma_2 J_l \tag{7.15}$$

where γ_1 and γ_2 are positive weight coefficients that show how important one target is in relation to the other. Therefore, the control strategy may be developed by minimizing the cost function (7.15) while meeting the constraints (7.8), (7.9) and (7.11), which can be transformed into the following constrained optimization problem:

$$\begin{aligned} &\underset{u_1(k), u_2(k)}{\text{minimize}} \quad J(x(k+1), u_1(k), u_2(k)) \\ &\text{subject to} \quad x(k+1) = x(k) + d(C_1 u_1(k) + C_2 u_2(k) - b(k)) \\ &I_{B_c} 1_{mn} + b(k) \le C_1 u_1(k) + C_2 u_2(k) \le I_{B_d} 1_{mn} + b(k) \\ &0_{m-1} \le u_1(k) \le I_{me_u} 1_{m-1}, \; x_l 1_{mn} \le x(k+1) \le x_u 1_{mn} \\ &0_{m(n-1)} \le u_2(k) \le I_{ce_u} 1_{m(n-1)}. \end{aligned} \tag{7.16}$$

The majority of the current practices and tactics, including those found in [4, 16, 17], centre on decentralized equalizing control, where each equalizer is autonomously regulated to equalize its two neighboring cells. In contrast to these, the developed cell balancing approach in (7.16) globally optimizes all equalizing currents to accomplish the equalization of the cells' SOC while taking into account all relevant data and the limits placed on the cells and equalizers in the cell equalizing system. By synchronizing the regulation of various equalizers, this can have the benefits of reducing the time required for cell equilibration while also guaranteeing the cells' existing constraints. The number of optimization variables, however, is $mn-1$ in (7.16), which puts a heavy computational strain on the controller and makes it challenging to execute this cell equalizing approach in real-time for large-scale battery packs.

Here, a novel hierarchical optimum cell equalizing control technique is suggested as a means of addressing this shortcoming. It has the following two layers:

- The top layer is the module-level equalizing control, where each battery module's cell-level equalizing currents are set to zero and the battery modules are handled as independent cells. It is possible to create the module-level controlled equalizing currents $u_1(k)$ according to (7.16).
- The bottom layer then computes the regulated cell-level equalizing currents $u_{2,i}(k)$ $(1 \leq i \leq m)$ for various battery modules in parallel using the pre-designed module-level equalizing currents.

The computational cost of the cell equalizing control method may be greatly decreased by adopting this hierarchical architecture. The detailed implementation of the aforementioned hierarchical cell equalizing control approach will be discussed in the paragraphs that follow.

7.2.2 Top Layer: Module-Level Equalizing Control

Due to their serial connection design, the cells in each battery module within this module-based cell equalizing system get the identical equalizing current from the module-level equalizers. As a result, at the module-level, each battery module may be thought of as a separate cell. Only the module-level equalizers are regulated in the top-layer control, and the battery modules' cell-level equalizing currents are set at $u_2(k) = 0_{m(n-1)}$. The cell equalizing system model may therefore be rewritten as follows at the module level:

$$x^m(k+1) = x^m(k) + d(C_1^m u_1(k) - b^m(k)) \tag{7.17}$$

where $x^m(k) \triangleq [SOC_{m_1}(k), \cdots, SOC_{m_m}(k)]^T \in \mathbb{R}^m$ is the battery module's SOC vector, $b^m(k) = [I_s(k), \cdots, I_s(k)]^T \in \mathbb{R}^m$, ${C^m}_1 \in \mathbb{R}^{m\times(m-1)}$ is

$$C_1^m = \begin{bmatrix} -p_{m_1} & 0 & \cdots & 0 \\ p'_{m_1} & -p_{m_2} & \cdots & 0 \\ \cdots & \cdots & \cdots & \cdots \\ 0 & 0 & \cdots & p'_{m_{m-1}} \end{bmatrix}$$

The best regulated module-level equalizing currents may be estimated by resolving the following restricted optimization programming issue, similar to (7.16), based on the model in (7.17).

$$\begin{aligned} &\underset{u_1(k)}{\text{minimize}} \;\; J_m(x^m(k+1), u_1(k)) \\ &\text{subject to} \\ &x^m(k+1) = x^m(k)+d(C^m{}_1\, u_1(k)-b^m(k)) \\ &I_{B_c}1_m+b^m(k) \leq C_1^m u_1(k) \leq I_{B_d}1_m+b^m(k) \\ &0_{m-1} \leq u_1(k) \leq I_{me_u}1_{m-1},\;\; x_l 1_m \leq x^m(k+1) \leq x_u 1_m \end{aligned} \tag{7.18}$$

with

$$J_m = \gamma_1(x^m(k+1)\text{-}\bar{x}^m(k+1))^T(x^m(k+1)\text{-}\bar{x}^m(k+1))+\gamma_2 u_1^T(k)u_1(k)$$

where $\bar{x}^m(k) = \frac{1}{m}1_m 1_m^T x^m(k)$. By introducing a barrier function [18] to replace the inequality constraints, the constrained optimization problem (7.18) can be transformed as

$$\underset{u_1(k)}{\text{minimize}}\; J_m(x^m(k+1), u_1(k)) - \frac{1}{\mu}\sum_{i=1}^{6m-2}\log(-g_i(u_1(k))) \tag{7.19}$$

where the second item in (7.19) denotes the logarithmic barrier function, μ is a positive parameter, and $g_i(u_1(k))$ is the i-th element of the vector $g(u_1(k)) \in \mathbb{R}^{6m-2}$ given as

$$g(u_1(k)) = \begin{bmatrix} C_1^m u_1(k) - b^m(k) - I_{B_d}1_m \\ -C_1^m u_1(k) + b^m(k) + I_{B_c}1_m \\ d(C_1^m u_1(k) - b^m(k)) + x^m(k) - x_u 1_m \\ -d(C_1^m u_1(k) - b^m(k)) - x^m(k) + x_l 1_m \\ u_1(k) - I_{me_u}1_{m-1} \\ -u_1(k) \end{bmatrix}. \tag{7.20}$$

The optimization issue (7.19) is solved using the interior point based technique [18], allowing the calculation of the ideal module-level control equalizing currents $u_1^*(k)$.

7.2.3 Bottom Layer: Cell-Level Equalizing Control

The cell-level equalizing currents are constructed in the bottom layer, where each battery module's computation is done concurrently. With the pre-designed module-level equalizing currents $u_1^*(k)$ by the top layer, the equalizing model for the i-th $(1 \leq i \leq m)$ battery module may be expressed as

$$x_i(k+1) = x_i(k) + d(1_n C_{1,i}^m u_1^*(k) + C_{2,i} u_{2,i}(k) - b_i(k)) \tag{7.21}$$

where $C_{1,i}^m \in \mathbb{R}^{1\times(m-1)}$ is the i-th row of C_1^m in (7.17), $C_{2,i}$, $u_{2,i}(k)$, and $b_i(k)$, and $x_i(k)$ are the SOC vectors for the cells in the i-th battery module (7.7). The cell-level equalizing control method is created for the cells in the i-th battery module based on (7.21) as:

$$\begin{aligned}
&\underset{u_{2,i}(k)}{\text{minimize}} \;\; J_{c_i}(x_i(k+1), u_{2,i}(k)) \\
&\text{subject to} \\
&x_i(k+1) = x_i(k) + d(1_n C_{1,i}^m u_1^*(k) + C_{2,i} u_{2,i}(k) - b_i(k)) \\
&I_{B_c} 1_n \leq C_{2,i} u_{2,i}(k) + 1_n C_{1,i}^m u_1^*(k) - b_i(k) \leq I_{B_d} 1_n \\
&0_{n-1} \leq u_{2,i}(k) \leq I_{ce_u} 1_{n-1}, \; x_l 1_n \leq x_i(k+1) \leq x_u 1_n
\end{aligned} \tag{7.22}$$

with

$$J_{c_i} = \gamma_1 (x_i(k+1) - \bar{x}_i(k+1))^T (x_i(k+1) - \bar{x}_i(k+1)) + \gamma_2 u_{2,i}^T(k) u_{2,i}(k)$$

where $\bar{x}_i(k) = \frac{1}{n} 1_n 1_n^T x_i(k)$. The interior point based approach may also be used to determine optimum $u_{2,i}^*(k)$ for $1 \leq i \leq m$ [18]. The ideal regulated cell-level equalizing currents are then discovered to be $u_2^*(k) = [u_{2,1}^*(k), \cdots, u_{2,m}^*(k)]^T$.

It is found that the top-layer control algorithm's (7.18) optimization variables have a dimension of $m-1$ and that the bottom layer's m parallel optimization vectors have a size of $n-1$. The computing cost of the hierarchical cell equalizing control approach on the controller is substantially lower than the linked $mn-1$ optimization variables that must be addressed in (7.18), making cell equalization implementation in real-time more practical.

Convergence analysis: The suggested cell equalizing method's mathematical convergence property is encapsulated in the following theorem.

Theorem 7.1 *For the module-based cell equalizing system, the cells' SOCs can converge to their average value with the optimal cell equalizing currents $u_1^*(k)$ and $u_2^*(k)$ designed by the hierarchical cell equalizing control method.*

Proof The proving process may be broken down into two phases. First, each battery module's SOC convergence evidence is given. Then, it is demonstrated that the SOCs of every cell in the battery pack converge to their average value. A Lyapunov candidate function for the i-th $(1 \leq i \leq m)$ battery module is defined as follows:

$$V_i(k) = (x_i(k) - \bar{x}_i(k))^T (x_i(k) - \bar{x}_i(k)). \tag{7.23}$$

It can be seen from (7.23) that $V_i(k) \geq 0$, and $V_i(k) = 0$ if and only if $x_i(k) = \bar{x}_i(k)$. According to (7.22), $u_{2,i}^*(k)$ minimizes J_{c_i} at each sampling step since it is a convex optimization, which may be stated as follows:

$$J_{c_i}(x_i(k+1), u_{2,i}^*(k)) \leq J_{c_i}(x_i(k+1), u_{2,i}(k)) \tag{7.24}$$

for all $u_{2,i}(k)$ that satisfies the constraints in (7.22). Since $u_{2,i}(k)= 0_{n-1}$ is a feasible solution, it yields

$$J_{c_i}(x_i(k+1), u_{2,i}^*(k)) \leq J_{c_i}(x_i(k+1), 0_{n-1}). \tag{7.25}$$

If and only if $u_{2,i}^*(k) = 0_{n-1}$, which signifies the circumstance when all cells' SOCs in the i-th battery module converge to the same value and the cell equalizing procedure is completed, the left side and right side of the equation in (7.25) are equal. Inferring from (7.22), it may be said that

$$J_{c_i}(x_i(k+1), 0_{n-1}) = \gamma_1 (x_i(k) - \bar{x}_i(k))^T (x_i(k) - \bar{x}_i(k)). \tag{7.26}$$

By substituting (7.26) into (7.25), for the Lyapunov function (7.23), the following inequality can be obtained

$$V_i(k+1) - V_i(k) \leq -\tfrac{\gamma_2}{\gamma_1} u_{2,i}^{*T}(k) u_{2,i}^*(k). \tag{7.27}$$

Only when $u_{2,i}^*(k) = 0_{n-1}$ does the right side of (7.27) have a positive value and equal to zero. The LaSalle's invariance principle [19] leads to the conclusion that $x_i(k) \to \bar{x}_i(k)$ for $1 \leq i \leq m$ is true. The i-th element of $x^m(k)$ is therefore determined by $x(k) \to [x_1^m(k)1_n^T, \cdots, x_m^m(k)1_n^T]^T$ with $x_i^m(k)$.

Similarly, it can be demonstrated that $x^m(k) \to \bar{x}^m(k)$ since $u_1^*(k)$ minimizes J_m at each sampling step based on (7.18). So, it is demonstrable that $x(k) \to \bar{x}(k)$. Thus, it is demonstrated that the suggested hierarchical optimum control method and the closed-loop cell equalizing system are convergent.

7.3 Results and Discussions

Experiments are performed on a battery pack with a nominal capacity and voltage of 3.1 Ah, and 44.4 V in order to confirm the effectiveness of the hierarchical optimum cell equalizing control method suggested above. It consists of $n = 4$ serially linked cells with a nominal capacity of 3.1 Ah and a nominal voltage of 3.7 V, distributed across $m = 3$ battery modules. The reference [20] contains the parameters governing the temperature dynamics of the cells. The cells' SOCs have maximum and lower

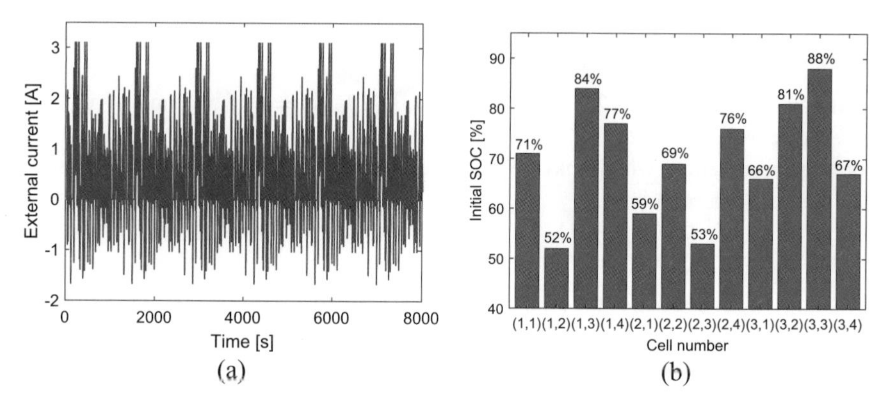

Fig. 7.3 **a** External current for the battery pack, **b** initial SOCs of the cells

limits of 100% and 0%, respectively. Cell-level and module-level equalizers are both permitted to operate at a maximum equalizing current and energy transfer efficiency of 1 A and 0.9 A, respectively. The currents in the cells need to be kept between [−3.1 A, 3.1 A] (1-C rate). Fig. 7.3a, which shows a series of FUDS [21] at a 1-C rate, illustrates the external current of the battery pack. The sample time is $T = 1$ s, and the weight coefficients are accordingly set at $\gamma_1 = 10^4$, $\gamma_2 = 1$. The equalizing time is the amount of time consumed when the RMS of the cells' SOC differences reaches an acceptable bound of $\frac{1}{\sqrt{mn}}||x(k)\text{-}\bar{x}(k)|| \le 0.25\%$, where $||\cdot||$ stands for the 2-norm. Setting $u_1(k) = 0_{m-1}$ and $u_2(k) = 0_{m(n-1)}$, at this point terminates the cell equalizing operation. This can successfully prevent over equalization and lower the cost of energy supply for the equalizer circuits. Keep in mind that alternative appropriate values may also be provided, based on the demands and effectiveness in practice.

7.3.1 Cell Equalizing Results

The starting SOCs of the cells are chosen at random as seen in Fig. 7.3a, with an RMS variation in SOC of 11.13%. In Fig. 7.4a–d, we show the cell equalizing outcomes after using our suggested technique in terms of the cells' SOCs, the disparities between the cells' SOCs and their average value, the cells' temperatures, and the cells' currents. This shows the superior performance of the suggested hierarchical optimum cell equalizing strategy as the cells' SOC may converge to their average value with the equalizing duration of 2277 s and the cells' maximum temperature of 32.84°C. Even when the battery pack is operating with a significant external current, the cells' currents within their narrow range of [−3.1 A, 3.1 A] may be well assured. Figure 7.5a–d, respectively, depicts the corresponding regulated module-level and cell-level equalizing currents. It demonstrates that the regulated equalizing currents

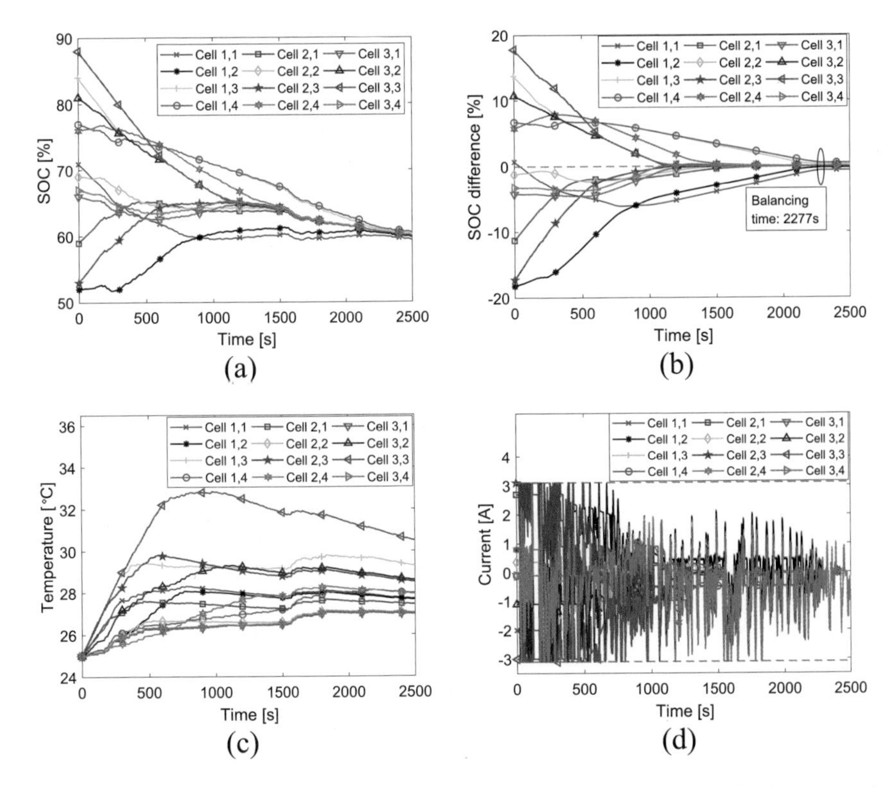

Fig. 7.4 **a** SOCs, **b** SOC differences, **c** temperatures, **d** currents of the cells under the proposed cell equalizing control

fluctuate in response to changes in the external current, thereby preventing any potential violations of the current restrictions of the cells, which is consistent with the study shown above. Figure 7.6 displays the SOC responses and currents of the cells under the ideal equalizing control (7.16). It equalizes in 2249 s, which is comparable to the time required by the suggested hierarchical optimum cell equalizing method. It demonstrates how the hierarchical control structure may lessen the computational load at the cost of a slight increase in equalization time.

7.3.2 *Tests of Different Weight Selections*

The weight factors in the cost functions (7.18) and (7.22) describe the relative weights of the objectives in terms of the rate of temperature rise and cell equalization, respectively. More experiments with various weight coefficients are built, where $\gamma_1 = 10^4$ is fixed and γ_2 is chosen as 0.05, 0.1, 0.5, 1, 5, and 10, respectively, to examine the impact of weight selection on the cell equalizing performance. Figure 7.7 displays

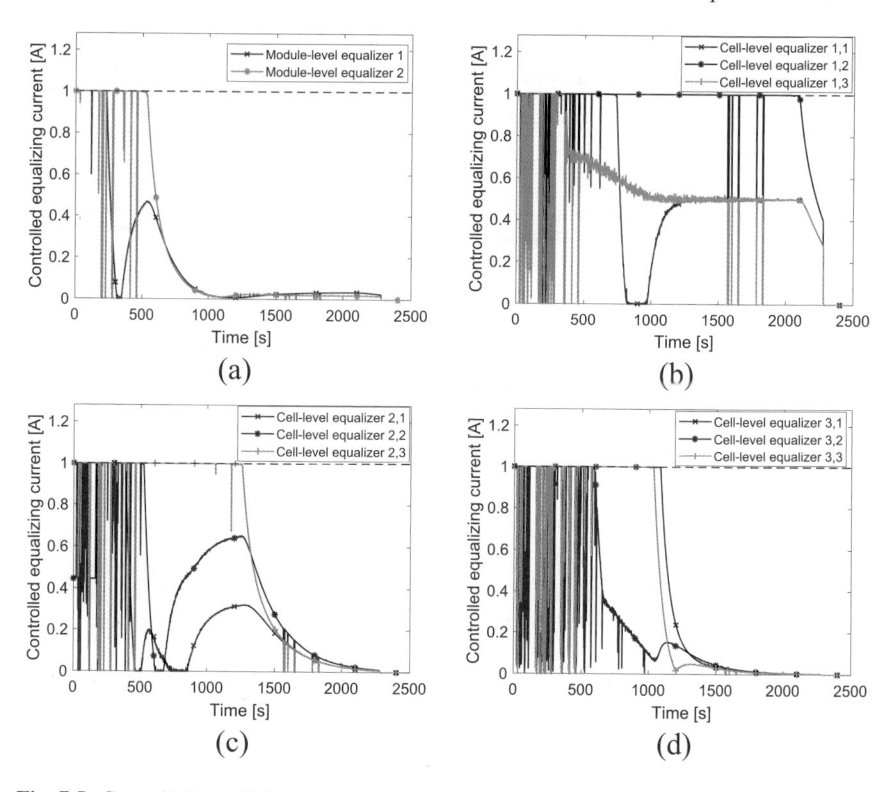

Fig. 7.5 Controlled equalizing currents under the proposed cell equalizing control of **a** module-level equalizers, **b** cell-level equalizers in battery module 1, **c** cell-level equalizers in battery module 2, **d** cell-level equalizers in battery module 3

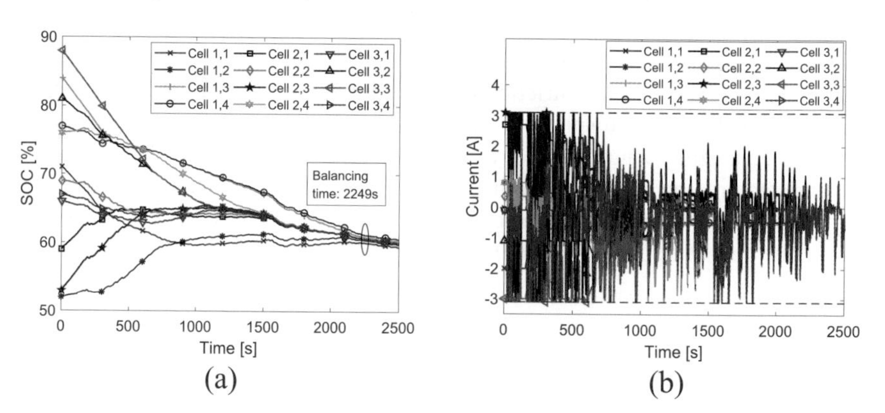

Fig. 7.6 **a** SOCs, **b** currents of the cells under the optimal cell equalizing control (7.16)

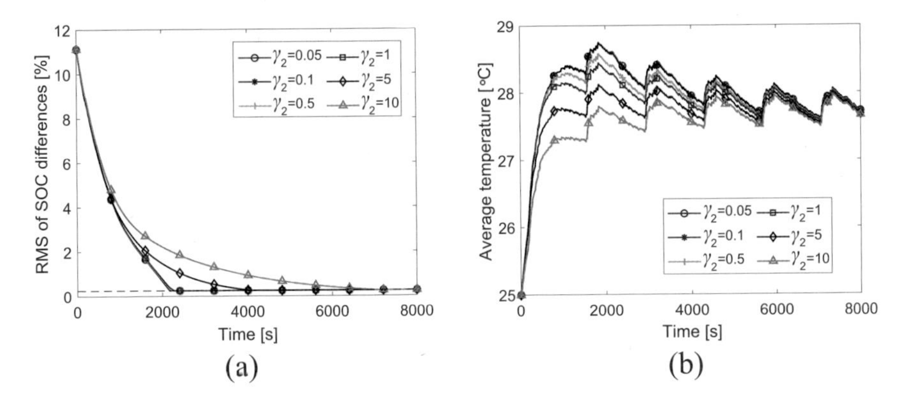

Fig. 7.7 **a** RMS of the cells' SOC differences, **b** cells' average temperatures for different weight coefficients

the results of the RMS of the differences in SOC between the cells and the average temperatures of the cells. It is obvious that a higher γ_2 would result in a gradual increase in temperature but would negatively impact the performance of equalizing speed. In order to balance these objectives in practice, users can pick the appropriate weights using this rule as a reference. Based on the findings in Fig. 7.7, we chose $\gamma_2 = 1$ to build our cell equalizing control technique because it offers an acceptable trade-off between these two conflicting objectives.

7.3.3 Comparison With Decentralized Equalizing Control

The outcomes of the decentralized technique are also supplied as a comparison to show the superior performance of the developed hierarchical optimum cell equalizing control approach. Each equalizer that balances its two neighboring cells is separately controlled in the decentralized cell balancing control scheme. A cautious procedure similar to [4] that substitutes $\frac{1}{2}(I_{B_c} + I_s(k)) \leq I_{me_i}(k) \leq \frac{1}{2}(I_{B_d} + I_s(k))$ and $\frac{1}{2}(I_{B_d} + I_s(k) - C^m_{1,i}u^*_1(k)) \leq I_{ce_{i,j}}(k) \leq \frac{1}{2}(I_{B_c} + I_s(k) - C^m_{1,i}u^*_1(k))$ for the current restrictions in (7.18) and (7.22) to decouple the relationship between equalizers is required to avoid the cells' currents from exceeding their limits. The regulation of each module-level and cell-level equalizer may thus be accomplished using decentralized algorithms that are based on (7.18) and (7.22) and take into account just the equalization of two nearby cells. Figure 7.8 illustrates the SOC responses and currents of the cells under the decentralized equalizing control, with an equalizing duration of 2416 s and well-guaranteed current constraints for the cells. The suggested optimum cell equalization technique may reduce the equalizing time by 139 s when compared to the decentralized equalizing control approach, according to the comparative results of Figs. 7.4 and 7.8. It supports the suggested cell equalizing strategy's promising performance.

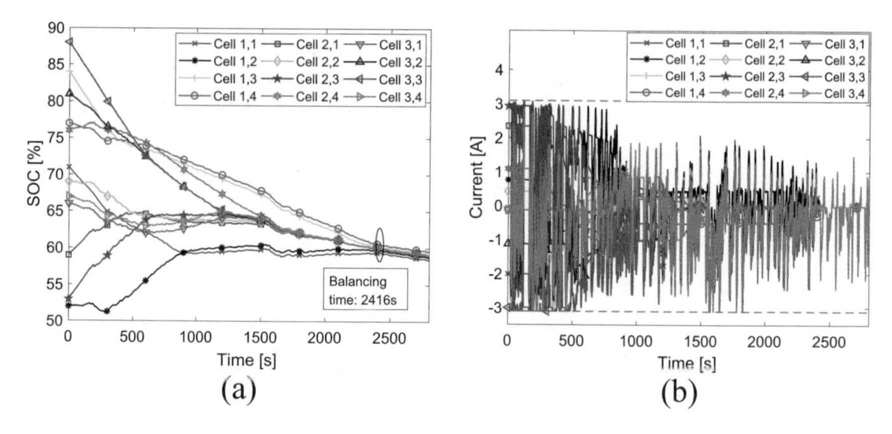

Fig. 7.8 **a** SOCs, **b** currents of the cells under the decentralized equalizing control

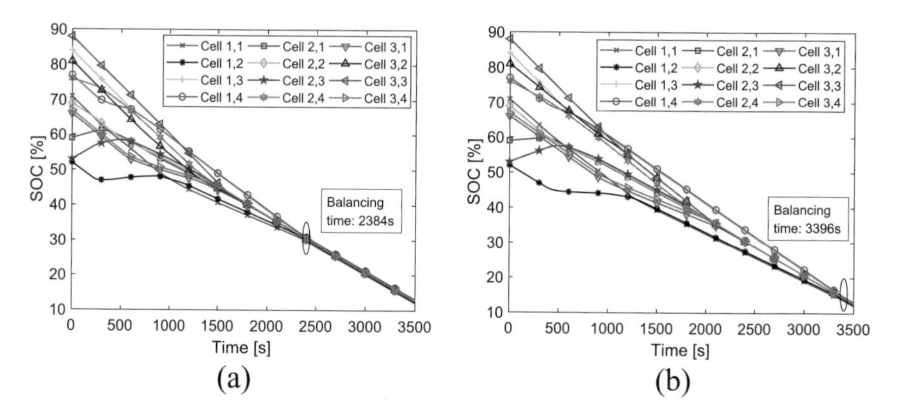

Fig. 7.9 **a** SOCs under the proposed cell equalizing control, **b** SOCs under the decentralized equalizing control

The outcomes of the suggested hierarchical optimum and decentralized cell equalizing control systems are shown in Fig. 7.9 for the scenario where the battery pack operates with a constant discharging current of 1.8 A, where the equalizing periods are 3396 s and 2384 s, respectively. According to the comparison, the equalizing time is decreased by 29.8% when the coordinated control of all equalizers is taken into account, which is consistent with the study presented above.

7.3.4 Tests for Different Cells' Initial SOCs

In the last, further experiments are conducted with the external current shown in Fig. 7.3a, which enables cells to start from varied SOCs as demonstrated in case 1–4

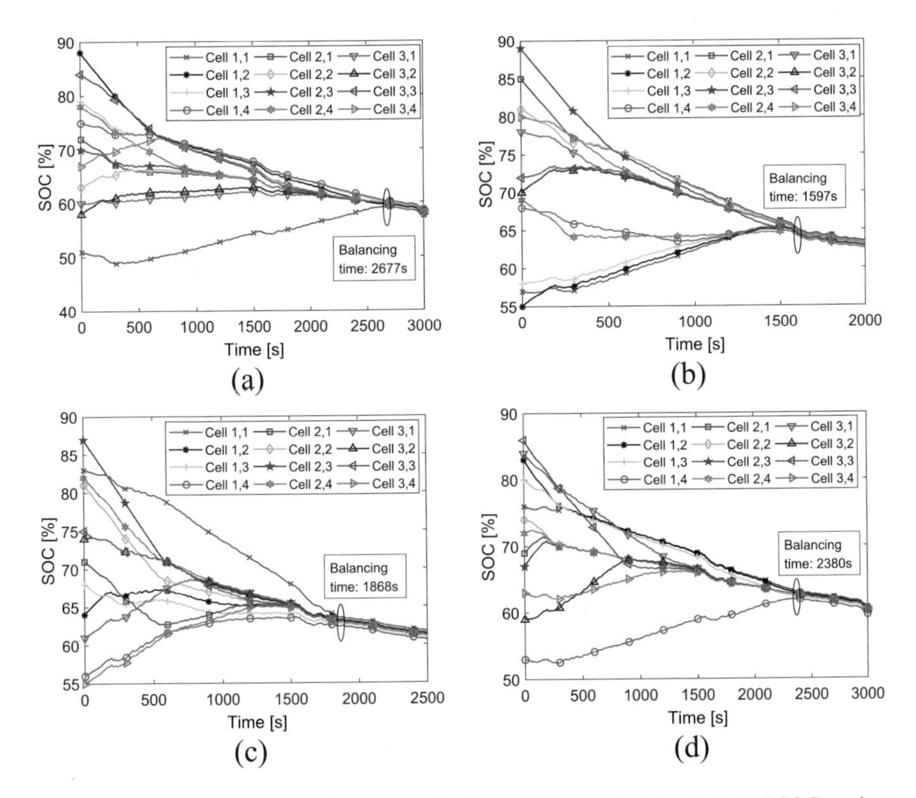

Fig. 7.10 SOC responses under the proposed cell equalizing control for the initial SOCs as in **a** Case 1, **b** Case 2, **c** Case 3, **d** Case 4

to balance, in order to investigate the performance of the proposed cell equalizing approach for different beginning cells' SOCs.

- Case 1: $x(0) = [51\%, 88\%, 79\%, 75\%, 72\%, 63\%, 70\%, 78\%, 60\%, 58\%, 84\%, 67\%]^T$;
- Case 2: $x(0) = [57\%, 55\%, 58\%, 68\%, 85\%, 81\%, 89\%, 69\%, 78\%, 70\%, 72\%, 80\%]^T$;
- Case 3: $x(0) = [83\%, 64\%, 68\%, 56\%, 71\%, 81\%, 87\%, 82\%, 61\%, 74\%, 75\%, 55\%]^T$;
- Case 4: $x(0) = [76\%, 83\%, 80\%, 53\%, 69\%, 74\%, 67\%, 72\%, 84\%, 59\%, 86\%, 63\%]^T$.

The outcomes of the cells' SOC responses to these initial SOCs are shown in Fig. 7.10a–d. It illustrates that the planned hierarchical optimum equalizing approach performs satisfactorily by showing that the cells' cell equalization can be achieved in all circumstances.

References

1. C.H. Kim, M.Y. Kim, H.S. Park, G.W. Moon, A modularized two-stage charge equalizer with cell selection switches for series-connected lithium-ion battery string in an HEV. IEEE Trans. Power Electron. **27**(8), 3764–3774 (2012)
2. H. Chen, L. Zhang, Y. Han, System-theoretic analysis of a class of battery equalization systems: mathematical modeling and performance evaluation. IEEE Trans. Veh. Technol. **64**(4), 1445–1457 (2015)
3. J. Gallardo-Lozano, E. Romero-Cadaval, M.I. Milanes-Montero, M.A. Guerrero-Martinez, Battery equalization active methods. J. Power Sources **246**, 934–949 (2014)
4. Q. Ouyang, J. Chen, J. Zheng, Y. Hong, Soc estimation-based quasi-sliding mode control for cell balancing in lithium-ion battery packs. IEEE Trans. Ind. Electron. **65**(4), 3427–3436 (2018)
5. Y.S. Lee, M.W. Cheng, Intelligent control battery equalization for series connected lithium-ion battery strings. IEEE Trans. Ind. Electron. **52**(5), 1297–1307 (2005)
6. W. Han, C. Zou, C. Zhou, L. Zhang, Estimation of cell soc evolution and system performance in module-based battery charge equalization systems. IEEE Trans. Smart Grid. **10**(5), 4717–4728 (2019)
7. Q. Ouyang, J. Chen, J. Zheng, H. Fang, Optimal cell-to-cell balancing topology design for serially connected lithium-ion battery packs. IEEE Trans. Sustain. Energy **9**(1), 350–360 (2018)
8. H. Fang, X. Zhao, Y. Wang, Z. Sahinoglu, T. Wada, S. Hara, R.A. de Callafon, Improved adaptive state-of-charge estimation for batteries using a multi-model approach. J. Power Sources **254**, 258–267 (2014)
9. J. Wei, G. Dong, Z. Chen, Lyapunov-based state of charge diagnosis and health prognosis for lithium-ion batteries. J. Power Sources **397**, 352–360 (2018)
10. J. Chen, Q. Ouyang, C. Xu, H. Su, Neural network-based state of charge observer design for lithium-ion batteries. IEEE Trans. Control. Syst. Technol. **26**(1), 313–320 (2017)
11. C. Zou, X. Hu, S. Dey, L. Zhang, X. Tang, Nonlinear fractional-order estimator with guaranteed robustness and stability for lithium-ion batteries. IEEE Trans. Ind. Electron. **65**(7), 5951–5961 (2018)
12. L. Yuang-Shung, D. Jiun-Yi, Fuzzy-controlled individual-cell equaliser using discontinuous inductor current-mode Cuk convertor for lithium-ion chemistries. IEE Proc.-Electr. Power Appl. **152**(5), 1271–1282 (2005)
13. J. Yan, Z. Cheng, G. Xu, H. Qian, Y. Xu, Fuzzy control for battery equalization based on state of charge, in *IEEE 72nd Vehicular Technology Conference-Fall*, (Ottawa, ON, Canada, 2010), pp. 1–7
14. M.F. Samadi, M. Saif, Nonlinear model predictive control for cell balancing in li-ion battery packs, in *American Control Conference*, (Portland, OR, USA, 2014), pp. 2924–2929
15. Q. Ouyang, J. Chen, J. Zheng, H. Fang, Optimal multiobjective charging for lithium-ion battery packs: a hierarchical control approach. IEEE Trans. Ind. Inform. **14**(9), 4243–4253 (2018)
16. M. Caspar, T. Eiler, S. Hohmann, Systematic comparison of active balancing: a model-based quantitative analysis. IEEE Trans. Veh. Technol. **67**(2), 920–934 (2018)
17. M.-Y. Kim, J.-H. Kim, G.-W. Moon, Center-cell concentration structure of a cell-to-cell balancing circuit with a reduced number of switches. IEEE Trans. Power Electron. **29**(10), 5285–5297 (2014)
18. S. Boyd, S.P. Boyd, L. Vandenberghe, *Convex Optimization*, (Cambridge university press, 2004)
19. H.K. Khalil, *Nonlinear Systems*, (Prentice Hall, Englewood Cliffs, NJ, USA, 2002)
20. A. Abdollahi, X. Han, G. Avvari, N. Raghunathan, B. Balasingam, K.R. Pattipati, Y. Bar-Shalom, Optimal battery charging, part i: minimizing time-to-charge, energy loss, and temperature rise for ocv-resistance battery model. J. Power Sources **303**, 388–398 (2016)
21. H. He, R. Xiong, X. Zhang, F. Sun, J. Fan, State-of-charge estimation of the lithium-ion battery using an adaptive extended kalman filter based on an improved thevenin model. IEEE Trans. Veh. Technol. **60**(4), 1461–1469 (2011)

Chapter 8
Module-Based Cell-to-Pack Equalization Control

8.1 Improved Module-Based CPC Equalization System

The CPC equalizing system [1–5] is one of the most widely and commonly used equalization systems in practice, where the equalizers are controlled to transfer energy between each cell and the whole battery pack to accomplish the cell equalization, as illustrated in Fig. 8.1. However, the CPC structure is not appropriate for a battery pack made up of a large number of cells connected in series for the following two reasons:

1. The increased number of cells raises the total voltage on the battery pack side, putting a significant strain on the circuit component of the equalizers. In practice, it is nearly impossible to directly implement for large-scale battery packs. As an example, a single LTC3300-1 board can only balance six cells with a total voltage of the battery pack (module) side of up to 36 V [1].
2. The traditional CPC structure is not modular. The equalization system for a large-scale battery pack in EV applications would be very complex and massive, posing additional challenges for system repair and maintenance.

To address these shortcomings, an enhanced module-based CPC equalizing system has been developed here, including its structure shown in Fig. 8.2. The battery pack is partitioned into m battery modules, each with n battery cells. When the cell's SOC is larger/smaller than the average one of the battery modules, the CMC equalizer is used within each module to move energy from the cell/battery module to the battery module/cell, where the cells' SOCs can be assumed to be known by using well-studied SOC estimation techniques [6, 7]. The module-to-module balancers are used as ML equalizers for energy equalization between adjacent battery modules, which are similar to cell-to-cell equalizers in that each battery module is treated as a high-voltage individual cell. This operation can greatly reduce the challenge to the withstand voltage of the electrical components in the equalizing system because the voltage of each battery module is much lower than that of the entire battery pack. Furthermore, the simple module-to-module structure makes the ML equalizers easily

J. Chen et al., *Equalization Control for Lithium-Ion Batteries*,
https://doi.org/10.1007/978-981-99-0220-0_8

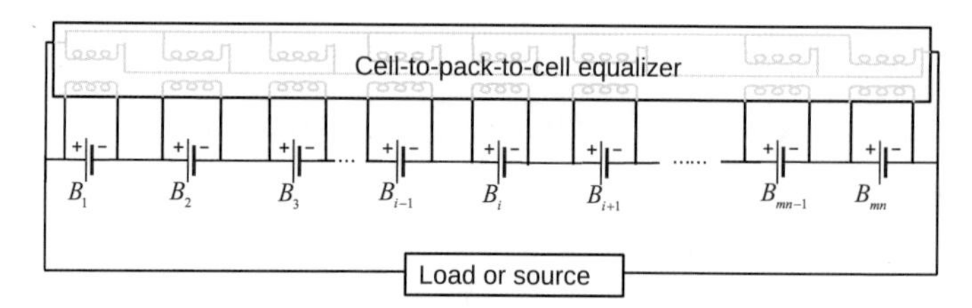

Fig. 8.1 Diagram of the conventional CPC equalization system

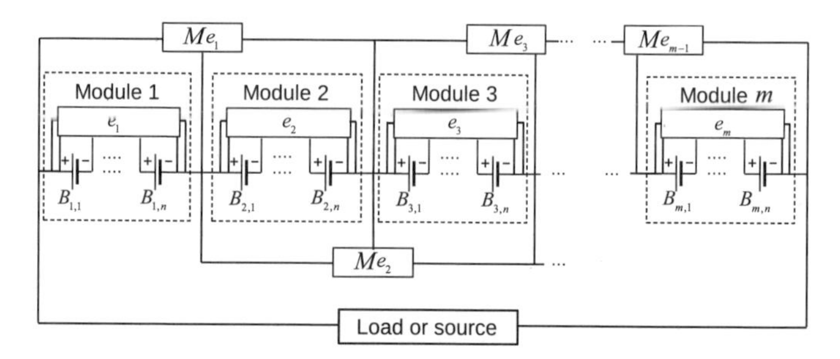

Fig. 8.2 Diagram of the improved module-based CPC equalization system

replaceable in the equalization system and the battery modules, making battery pack and equalization system maintenance much more convenient than the conventional CPC equalizing system. It demonstrates that the designed enhanced module-based CPC equalization system can fully exploit the benefits of the CPC structure and module-to-module structure to achieve enhanced equalization performance.

The bidirectional isolated modified buck-boost converters [1] are used as CMC equalizers in this work, and the modified bidirectional Cûk converters [7] are used as ML balancers. We only intend to design the corresponding equalizing current control strategy, assuming that the equalizing current can be appropriately supplied, because they are already well controlled in previous work.

8.1.1 Equalizing Current Formulation

As depicted in Fig. 8.2, the total equalizing current is the current through its connected CMC equalizer e_i and ML equalizers Me_{i-1} and Me_i in each cell of the i-th ($2 \leq i \leq m-1$) battery module. The first/m-th cells in the battery module receive the equalizing current only from the equalizers e_1/e_m and Me_1/Me_m.

CMC equalizing current: The CMC equalizer e_i consists of n equalizer modules, defined as $e_{i,j}$ ($1 \leq j \leq n$), which transfers energy independently between $B_{i,j}$-th cells and the i-th battery module in parallel. The cells' equalization can be achieved if the energy transfer takes place from $B_{i,j}$ to the i-th module with $SOC_{i,j}(k) > S\bar{O}C_i(k)$, where $SOC_{i,j}(k)$ is the SOC of $B_{i,j}$ and $S\bar{O}C_i(k)$ is the average SOC of the i-th battery module with $S\bar{O}C_i(k) = \frac{1}{n}\sum_{j=1}^{n} SOC_{i,j}(k)$. According to the law of conservation of energy, it can be calculated that

$$\alpha_{i,j} V_{B_{i,j}}(k) I_{cm_{i,j}}(k) = (\sum_{j=1}^{n} V_{B_{i,j}}(k)) I_{mc_{i,j}}(k) \quad (8.1)$$

where $V_{B_{i,j}}(k)$ is the terminal voltage of $B_{i,j}$, $\alpha_{i,j}$ is the energy transfer efficiency of the equalizer module $e_{i,j}$ when the cell $B_{i,j}$ charges the i-th battery module, $I_{cm_{i,j}}(k)/I_{mc_{i,j}}(k)$ is the CMC equalizing current of the cell/module side, respectively. Because the battery's internal resistance is low, the terminal voltages of the cells are supposed to be the same to simplify the problem complexity. Then, based on (8.1), it is possible to conclude that

$$I_{mc_{i,j}}(k) = \frac{\alpha_{i,j}}{n} I_{cm_{i,j}}(k). \quad (8.2)$$

It is to be noted that all the cells in the i-th battery module, including $B_{i,j}$ itself, receive the same equalizing current $I_{mc_{i,j}}(k)$ because they are parts of the battery module. When $SOC_{i,j}(k) < S\bar{O}C_i(k)$, the energy should be moved from the i-th module to $B_{i,j}$. Similarly, the relationship between the equalizing currents can be formulated as

$$I_{cm_{i,j}}(k) = \beta_{i,j} n I_{mc_{i,j}}(k) \quad (8.3)$$

where $\beta_{i,j}$ is the energy transfer efficiency of $e_{i,j}$ when the i-th module charges $B_{i,j}$.

To homogenize the abbreviations for both case scenarios where energy is being transferred from $B_{i,j}$/the i-th module to the i-th module/$B_{i,j}$, the direction of current out/into of $B_{i,j}$/the i-th module is defined as the reference direction of the CMC equalizing current in (8.4). In reference to (8.2)–(8.3), we define a new variable $I_{c_{i,j}}(k)$ which is utilized to characterize $I_{cm_{i,j}}(k)$ and $I_{mc_{i,j}}(k)$ in both the cases as follows:

$$\begin{aligned} I_{cm_{i,j}}(k) &= k_{i,j}\,\mathrm{sgn}(SOC_{i,j}(k) - S\bar{O}C_i(k)) I_{c_{i,j}}(k) \\ I_{mc_{i,j}}(k) &= k'_{i,j}\,\mathrm{sgn}(SOC_{i,j}(k) - S\bar{O}C_i(k)) I_{c_{i,j}}(k) \end{aligned} \quad (8.4)$$

with

$$\begin{cases} k_{i,j} = 1,\ k'_{i,j} = \frac{\alpha_{i,j}}{n}, & \text{if } SOC_{i,j}(k) \ge S\bar{O}C_i(k) \\ k_{i,j} = \beta_{i,j} n,\ k'_{i,j} = 1, & \text{if } SOC_{i,j}(k) < S\bar{O}C_i(k) \end{cases}$$

where $I_{c_{i,j}}(k)$ defines the controlled CMC equalizing current, which is a positive value; $\mathrm{sgn}(\cdot)$ is the sign function.

The total CMC equalizing current of $B_{i,j}$ in the i-th battery module includes not merely the equalizing current of its battery side $I_{cm_{i,j}}(k)$, but also the current of the module $I_{mc_{i,j}}(k)$ across all cells in the module, because $B_{i,j}$ is a part of the i-th module. In light of the previous analysis, it can be stated as

$$I_{ce_{i,j}}(k) = I_{cm_{i,j}}(k) - \sum_{j=1}^{n} I_{mc_{i,j}}(k) \tag{8.5}$$

where $I_{ce_{i,j}}(k)$ is the total CMC equalizing current of $B_{i,j}$, $I_{cm_{i,j}}(k)$ and $I_{mc_{i,j}}(k)$ can be replaced by (8.4).

ML equalizing current: To accomplish equalization among the battery modules, the ML equalizers Me_1,..., Me_{m-1} are used to transfer the energy from the higher average SOC battery module to the lower average SOC battery module. As shown in Fig. 8.2, Me_i $(1 \le i \le m-1)$ could indeed transfer energy among its adjacent modules i and $i+1$ using an equalizing current is expressed as

$$\begin{aligned} I_{ml_i}(k) &= k_{m_i} \operatorname{sgn}(S\bar{O}C_i(k) - S\bar{O}C_{i+1}(k)) I_{m_i}(k) \\ I_{mr_i}(k) &= k'_{m_i} \operatorname{sgn}(S\bar{O}C_i(k) - S\bar{O}C_{i+1}(k)) I_{m_i}(k) \end{aligned} \tag{8.6}$$

with

$$\begin{cases} k_{m_i} = 1, k'_{m_i} = \beta_{m_i}, \text{ if } S\bar{O}C_i(k) \ge S\bar{O}C_{i+1}(k) \\ k_{m_i} = \beta_{m_i}, k'_{m_i} = 1, \text{ if } S\bar{O}C_i(k) < S\bar{O}C_{i+1}(k) \end{cases}$$

where $I_{ml_i}(k)$ and $I_{mr_i}(k)$ define the i-th and (i+1)-th battery modules' equalizing currents through Me_i with direction of the current out/into of the battery module $i/i+1$ is defined as the reference direction; $I_{m_i}(k)$ and β_{m_i} are the controlled equalizing current and energy transfer efficiency of Me_i, respectively. Because of their serial connection, all cells in each battery module share the same ML equalizing current. The total ML equalizing current of $B_{i,j}$ $(1 \le i \le m)$, denoted as $I_{em_{i,j}}(k)$, can be calculated as

$$I_{em_{i,j}}(k) = -I_{mr_{i-1}}(k) + I_{ml_i}(k). \tag{8.7}$$

Note that $I_{mr_0}(k)$ and $I_{ml_m}(k)$ are always set as zeros in (8.7), since the first m-th battery module is only connected with the Me_1 / Me_{m-1}.

From Fig. 8.2 and the previous analysis, the total equalizing current for $B_{i,j}$ can be formulated as

$$I_{e_{i,j}}(k) = I_{ce_{i,j}}(k) + I_{em_{i,j}}(k) \tag{8.8}$$

where $I_{e_{i,j}}(k)$ is the total equalization current of $B_{i,j}$. By substituting (8.4)–(8.7) into (8.8), we obtain the following:

$$\begin{aligned} I_{e_{1,j}}(k) &= \gamma_{1,j} I_{c_{1,j}}(k) - \sum_{j=1}^{n} \gamma'_{1,j} I_{c_{1,j}}(k) + \gamma_{m_1} I_{m_1}(k) \\ I_{e_{i,j}}(k) &= \gamma_{i,j} I_{c_{i,j}}(k) - \sum_{j=1}^{n} \gamma'_{i,j} I_{c_{i,j}}(k) - \gamma'_{m_{i-1}} I_{m_{i-1}}(k) + \gamma_{m_i} I_{m_i}(k) \quad (2 \le i \le m-1) \\ I_{e_{m,j}}(k) &= \gamma_{m,j} I_{c_{m,j}}(k) - \sum_{j=1}^{n} \gamma'_{m,j} I_{c_{m,j}}(k) + \gamma'_{m} \quad I_{m_{m-1}}(k) \end{aligned} \tag{8.9}$$

with

$$[\gamma_{i,j} = k_{i,j}\text{sgn}(SOC_{i,j}(k) - S\bar{O}C_i(k))]$$

$$[\gamma'_{i,j} = k'_{i,j}\text{sgn}(SOC_{i,j}(k) - S\bar{O}C_i(k))]$$

$$[\gamma_{m_i} = k_{m_i}\text{sgn}(SOC_i(k) - S\bar{O}C_{i+1}(k))]$$

$$[\gamma'_{m_i} = k'_{m_i}\text{sgn}(S\bar{O}C_i(k) - S\bar{O}C_{i+1}(k))].$$

Note that $I_{c_{i,j}}(k)$ and $I_{m_i}(k)$ are auxiliary control inputs.

When $SOC_{i,j}(k) \geq S\bar{O}C_i(k)/S\bar{O}C_i(k) \geq S\bar{O}C_{i+1}(k)$, $I_{cm_{i,j}}(k)/I_{ml_i}(k)$ is the actual control variable for the CMC/module-to-module equalizer with the value set equal to the designed $I_{c_{i,j}}(k)/I_{m_i}(k)$. Otherwise, we can set the actual control variable $I_{mc_{i,j}}(k)/I_{mr_i}(k)$ equal to the designed $I_{c_{i,j}}(k)/I_{m_i}(k)$. The number of control variables in the equalization system can be cut in half by using auxiliary control inputs.

8.1.2 *Improved Module-Based CPC Equalization System Model*

Based on the Kirchhoff's law of current, the total current of $B_{i,j}$ $(1 \leq i \leq m,\ 1 \leq j \leq n)$, defined as $I_{B_{i,j}}(k)$, includes the battery pack's external working current $I_s(k)$ and the equalizing current $I_{e_{i,j}}(k)$ that can be expressed as follows:

$$I_{B_{i,j}}(k) = I_s(k) + I_{e_{i,j}}(k) \tag{8.10}$$

where $I_{B_{i,j}}(k)$ is positive or negative depends on the discharging or charging mode of $B_{i,j}$. In reference to [8, 9], the SOC of $B_{i,j}$ is defined as the rate of its remained capacity to the nominal capacity and can be formulated as follows:

$$SOC_{i,j}(k+1) = SOC_{i,j}(k) - d(I_s(k) + I_{e_{i,j}}(k)) \tag{8.11}$$

with $d = \frac{\eta_0 T}{Q}$, where Q is the cells' nominal capacity in the battery pack, T represents the sampling period, the variable η_0 defines the Coulombic efficiency. To simplify the notations, the state and control input vectors are defined as $x(k) = [x_1{}^T(k), \ldots, x_m{}^T(k)]^T \in \mathbb{R}^{mn}$ with $x_i(k) \triangleq [SOC_{i,1}(k), \ldots, SOC_{i,n}(k)]^T \in \mathbb{R}^n$ and $U(k) = [u_1^T(k), u_2^T(k)]^T \in \mathbb{R}^{mn+m-1}$ with $u_1(k) = [u_{1,1}^{T(k)}, \ldots, u_{1,m}^{T(k)}]^T \in \mathbb{R}^{mn}$, $u_{1,i}(k) \triangleq \left[I_{c_{i,1}}(k), \ldots, I_{c_{i,n}}(k)\right]^T \in \mathbb{R}^n$, and $u_2(k) \triangleq \left[I_{m1}(k), \ldots, I_{m_{m-1}}(k)\right]^T \in \mathbb{R}^{m-1}$ respectively. The improved module-based CPC equalizing model of the battery pack can then be expressed as the following state-space representation by substituting (8.9) into (8.11):

$$x(k+1) = x(k) + DU(k) - \phi(k) \tag{8.12}$$

where $\phi(k) \triangleq dI_s(k)1_{mn} \in \mathbb{R}^{mn}$, $D = \text{diag}\{D_1, D_2\} \in \mathbb{R}^{(mn+m-1)\times(mn+m-1)}$ with $D_1 = \text{diag}\{D_{1,1}, \ldots, D_{1,m}\} \in \mathbb{R}^{mn\times mn}$, $D_{1,i} = -d(\text{diag}\{\gamma_{i,1}, \ldots, \gamma_{i,n}\} - 1_n \times [\gamma'_{i,1}, \ldots, \gamma'_{i,n}]) \in \mathbb{R}^{n\times n}$, and $D_2 = {D^m}_2 \otimes 1_n \in \mathbb{R}^{mn\times(m-1)}$ with ${D^m}_2 \in \mathbb{R}^{m\times(m-1)}$ denoted as

$$D_2^m = d\begin{bmatrix} -\gamma_{m_1} & 0 & \cdots & 0 \\ \gamma'_{m_1} & -\gamma_{m_2} & \cdots & 0 \\ \cdots & \cdots & \cdots & \cdots \\ 0 & 0 & \cdots & \gamma'_{m_{m-1}} \end{bmatrix} \tag{8.13}$$

where $\text{diag}\{\cdot\}$ and $\otimes$ represent the diagonal matrix and the Kronecker product [10], respectively; 1_{mn} represents the column vector with mn ones.

It is to be noted that the enhanced module-based CPC equalization system model (8.12) is solely based on the SOC, irrespective of the voltage dynamics of the cells, which has at least two advantages. The first advantage is that, it broadens the model's applicability to different types of battery cells and equalization systems with that topology. The second advantage is that, without taking into account the cells' high-nonlinear voltage dynamics, the equalization control algorithm based on this model (8.12) is more straightforward to apply in practical applications.

8.2 Two-Layer Model Predictive Control Strategy

8.2.1 Cost Function Formulation

To accomplish cell equalization for the battery pack, the cells' SOCs need to regulate at the same level by driving the controlled CMC driving current and the controlled ML equalizing current in the enhanced module-based CPC equalization system, i.e., controlling $U(k)$ in (8.12) to satisfy the RMS of the cells' SOC difference as given below:

$$\frac{1}{\sqrt{mn}}||x(k) - \bar{x}(k)|| \le \epsilon \tag{8.14}$$

for any $x(0)$, where $||\cdot||$ defines the 2-norm, ϵ is the tolerance, $\bar{x}(k)$ represents the cells' average SOC vector that can be calculated as $\bar{x}(k) = \frac{1}{mn}1_{mn}1_{mn}^T x(k)$. As a result, the cost function of cell equalization can be formulated as follows by penalizing the bias between the cells' anticipated SOCs and their respective average value, as well as the control effort required [11]:

$$J(x(k)) = ||x(k+N)-\bar{x}(k+N)||_Q^2 + \sum_{i=0}^{N-1}(||x(k+i)-\bar{x}(k+i)||_Q^2 + ||U(k+i)||_R^2) \tag{8.15}$$

where $||x||_Q^2$ is the abbreviation of $x^T(k)Qx(k)$, N is the prediction horizon, Q and R are the corresponding positive weighting matrices, respectively.

8.2.2 Constraints

The safety of two main components, namely the equalizers and the cells, should be prioritized in the battery pack's enhanced module-based CPC equalizing system. As a result, the following safety-related hard constraints must be met during the cell equalization process.

CMC equalizing current restrictions: Because of the CMC equalizers' hardware limitations, the equalizing current should be less than the maximum value that the CMC equalizers could provide. It can be written as

$$0_{mn} \leq u_1(k) \leq U_{1u} 1_{mn} \tag{8.16}$$

with

$$\begin{cases} U_{1u} = I_{me_u}, & \text{if } \operatorname{sgn}(SOC_{i,j}(k) - S\bar{O}C_i(k)) \geq 0 \\ U_{1u} = \frac{1}{n} I_{me_u}, & \text{if } \operatorname{sgn}(SOC_{i,j}(k) - S\bar{O}C_i(k)) < 0 \end{cases}$$

where I_{me_u} defines the maximum allowed equalizing current in the cell side of the CMC equalizer, 0_{mn} represents the column vector with mn zeros. Because of the asymmetric configuration of the CMC equalizer, the bound U_{1u} fluctuates with the direction of the CMC equalizing current (the cell is controlled to charge or discharge the battery module).

ML equalizing current constraints: Correspondingly, due to physical constraints, the ML equalizing current should satisfy

$$0_{m-1} \leq u_2(k) \leq U_{2u} 1_{m-1} \tag{8.17}$$

where U_{2u} is the upper bound of the equalizing current that the ML equalizers can provide. The constraint of the input vector can be formulated by combining (8.16) and (8.17) as follows:

$$0_{mn+m-1} \leq U(k) \leq [\, U_{1u} 1_{mn}^T, U_{2u} 1_{m-1}^T]^T. \tag{8.18}$$

Cells' current limitations: Because a high current is detrimental and might cause permanent harm to the battery cells, the total current of the cell $I_{B_{i,j}}(k)$ $(1 \leq I \leq m, 1 \leq j \leq n)$ must be kept within the appropriate range as follows:

$$dU_{cm} 1_{mn} \leq DU(k) - \phi(k) \leq dU_{dm} 1_{mn} \tag{8.19}$$

where U_{cm} and U_{dm} are the cell's current boundary values, which demonstrate the maximum charging and discharging currents, respectively.

8.2.3 Centralized MPC Design

The MPC is a well-known model-based closed-loop optimization control technique that has demonstrated good performance in a variety of applications, including vehicles and smart grids. The MPC strategy can be an excellent alternative for cell equalization control issues if it is capable of dealing with problems with hard constraints, as mentioned by [12]. Using the enhanced module-based CPC equalization system model (8.12), the MPC-based equalizing control technique can be formulated by minimizing the cost function (8.15) while ensuring the hard constraints (8.18) and (8.19) as

$$\begin{aligned} &\underset{U(k),\ldots,U(k+N-1)}{\text{minimize}} \qquad J(x(k)) \\ &\text{subject to} \quad \forall i \in \{0, 1, \ldots, N-1\} \\ &x(k+i+1) = x(k+i) + DU(k+i) - \phi(k+i) \\ &dU_{cm}1_{mn} \leq DU(k+i) - \phi(k+i) \leq dU_{dm}1_{mn} \\ &0_{mn+m-1} \leq U(k+i) \leq [\, U_{1u}1^T_{mn}, U_{2u}1^T_{m-1}]^T. \end{aligned} \tag{8.20}$$

The optimal value of the control input vector $[U^{*T}(k), U^{*T}(k+1), \ldots, U^{*T}(k+N-1)]^T$ could be calculated by solving (8.20) at every sampling step k. With the help of receding horizon, the MPC converts the open-loop optimization problem (8.20) into a feedback control algorithm, which can improve the control strategy robustness. At this moment, it is to be noted that only the first step of the optimal solution[12], i.e., $U^*(k)$ can be implemented to the enhanced module-based CPC equalization system, however, the remaining optimal solutions $U^*(k+1), \ldots, U^*(k+N-1)$ are neglected.

Despite of the fact that the MPC technique (8.20) can resolve the cell equalization issues, there exists $N(mn+m-1)$ coupled control inputs in (8.20), that makes it difficult for the battery management controller to calculate (8.20) in real-time, especially in case of the battery pack comprising a large number of cells. A two-layer MPC technique is propounded to substitute the centralized control framework in (8.20), with a thorough explanation demonstrated in the following section, to minimize the computational burden of cell equalization control.

8.3 Two-Layer MPC for Cell Equalization

In Fig. 8.2, the enhanced module-based CPC equalization system is depicted which has been divided into two layers: the CMC-layer equalization and the module-layer equalization. If the controllers $u_2{}^*(k)$ and $u^*{}_{1,i}(k)$ $(1 \leq i \leq m)$ are designed separately for these two layers of equalization, the computational burden could be greatly

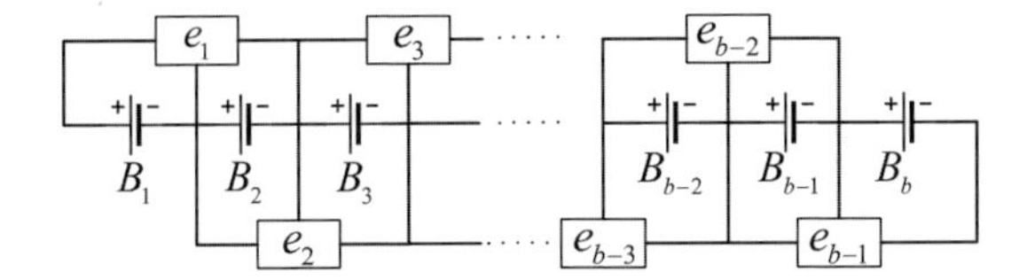

Fig. 8.3 Structure of the proposed two-layer MPC approach

reduced. In view of this motivation, a two-layer MPC-based equalizing control technique is propounded with its configuration depicted in Fig. 8.3.

- Although all cells inside every module share the same equalizing current provided by the ML equalizers, the top-layer MPC algorithm only designs the optimal controlled ML equalizing current $u_2{}^*(k)$, in which each battery module would be equivalent to a high-voltage individual cell.
- Throughout the bottom layer, the optimal controlled CMC equalizing current $u^*_{1,i}(k)$ $(1 \leq i \leq m)$ is determined by calculating for each battery module in parallel using the $u_2{}^*(k)$ pre-designed by the top-layer MPC.

The detailed top-layer and bottom-layer MPC strategies are described as in the following subsections.

8.3.1 Top-layer MPC: ML Equalizing Current Control

The ML equalizing current is the same for the cells inside every module, as illustrated in (8.2), because the cells are connected sequentially. Because of this property, evey battery module is regarded as a high-voltage individual cell in the top layer, with the CMC equalizing current set to $u_1(k) = 0_{mn}$. The battery pack can then be thought of as being made up of m cells, with the cell's SOC defined as the average SOC of each battery module in the original structure. With a new state vector of $z(k) \triangleq [S\bar{O}C_1(k), \cdots, S\bar{O}C_m(k)]^T \in \mathbb{R}^m$, the mathematical model of the ML equalization system can be calculated as

$$z(k+1) = z(k) + D_2^m\, u_2(k) - \phi^m(k) \tag{8.21}$$

with $\phi^m(k) \triangleq dI_s(k)1_m \in \mathbb{R}^m$, where D_2^m can be seen in (8.13). The top-layer MPC can be formulated similarly to (8.20) by optimizing the cost function in terms of the battery modules' SOC difference and control effort needed while assuring the respective hard constraints as follows:

$$
\begin{aligned}
\underset{u_2(k),\ldots,u_2(k+N-1)}{\text{minimize}} \quad & ||z(k+N)-\bar{z}(k+N)||^2_{Q_c} \\
& + \sum_{i=0}^{N-1}(||z(k+i)-\bar{z}(k+i)||^2_{Q_c} \\
& + ||u_2(k+i)||^2_{R_c}) \\
\text{subject to} \quad & \forall i \in \{0,1,\ldots,N-1\} \\
& z(k+i+1) = z(k+i) + D_2^m\, u_2(k+i) - \phi^m(k+i) \\
& dU_{cm}1_m \leq D_2^m\, u_2(k+i) - \phi^m(k+i) \leq dU_{dm}1_m \\
& 0_{m-1} \leq u_2(k+i) \leq U_{2u}1_{m-1}
\end{aligned} \tag{8.22}
$$

with $\bar{z}(k) = \frac{1}{m}1_m1_m^T z(k)$, where $[u_2^T(k), \ldots, u_2^T(k+N-1)]^T$ is the controlled ML equalizing current sequence denotes the variables that must be optimized at each sampling step, while k, Q_c and R_c denote the positive weighting matrices. The conventional approach proposed in [13] is very powerful method implemented to solve (8.22), and only the first step of the optimal solution $u_2(k) = u_2^*(k)$ is adopted while remaining optimal solutions $u_2^*(k+1), \ldots, u_2^*(k+N-1)$ are neglected.

8.3.2 Bottom-Layer MPC: CMC Equalizing Current Control

Because the top-layer MPC designs the optimal controlled ML equalizing current $u_2^*(k)$, it can be regarded as a known vector in the bottom layer. The cell equalizing model can be simplified by substituting $u_2(k) = u_2{}^*(k)$ into (8.12), and can be written as follows:

$$
x(k+1) = x(k) + D_1u_1(k) + D_2u_2{}^*(k) - \phi(k). \tag{8.23}
$$

It should be noted that the CMC equalizers only help to equalize the cells within each battery module. There is no interaction between modules. Consequently, the cell equalizing system (8.23) can be simplified by dividing it into m independent sub models as follows:

$$
x_i(k+1) = x_i(k) + D_{1,i}u_{1,i}(k) + 1_n D_{2_i}^m u_2^*(k) - \phi^c(k) \tag{8.24}
$$

for $1 \leq i \leq m$, where $D_{2_i}^m$ is the i-th row of D_2^m in (8.13), $\phi^c(k) \triangleq dI_s(k)1_n \in \mathbb{R}^n$. According to (8.24), the bottom-layer control technique could be composed of m decentralized MPC algorithms, that can be developed as

$$
\begin{aligned}
& \underset{u_{1,i}(k),\ldots,u_{1,i}(k+N-1)}{\text{minimize}} \quad ||x_i(k+N) - \bar{x}_i(k+N)||^2_{Q_i} \\
& \qquad\qquad + \sum_{l=0}^{N-1} (||x_i(k+l) - \bar{x}_i(k+l)||^2_{Q_i} \\
& \qquad\qquad + ||\, u_{1,i}(k+l)||^2_{R_i}) \\
& \text{subject to} \quad \forall l \in \{0, 1, \ldots, N-1\} \\
& x_i(k+l+1) = x_i(k+l) + D_{1,i}u_{1,i}(k+l) \\
& \qquad\qquad + 1_n D^m_{2_i} u^*_2(k) - \phi^c(k+l) \\
& dU_{cm} 1_n \leq D_{1,i}u_{1,i}(k+l) + 1_n D^m_{2_i}\, u^*_2(k) \\
& \qquad\qquad - \phi^c(k+l) \leq dU_{dm} 1_n \\
& 0_n \leq u_{1,i}(k+l) \leq U_{1u} 1_n
\end{aligned}
\tag{8.25}
$$

with $\bar{x}_i(k) = \frac{1}{n} 1_n 1_n^T x_i(k)$, Q_i and R_i the weighting matrices, respectively. Similarly, (8.25) can be solved by the barrier method and $u_{1,i}(k) = u^*_{1,i}(k)$ can be obtained at each sampling step k. Note that the distributed MPC algorithms can be computed in parallel, which can greatly improve the computing speed.

8.3.3 *Computational Complexity Comparison With Centralized MPC*

In the propounded two-layer MPC technique, the top-layer calculates the optimal controlled ML equalizing current vector $u_2^*(k)$ and the bottom-layer is responsible for developing m distributed MPC algorithms in parallel to obtain the optimal CMC equalizing current vector $u^*_1(k) = [u^{*T}_{1,1}(k), \ldots, u^{*T}_{1,m}(k)]^T$. In the top-layer MPC algorithm, the dimension of the optimization variables is $N(m-1)$, and in the bottom layer of the algorithm, there are m parallel control input vectors to be optimized with a dimension of Nn. Table 8.1 shows that the computational burden of the two-layer MPC is much lower than that of the centralized MPC with $N(m + mn - 1)$ coupled optimization variables in (8.20). This exemplifies the advantages of the two-layer equalization technique, which divides the control problem of a multi-coupling variable system (8.20) into multiple independent systems with fewer variables, as shown in (8.22) and (8.25). Furthermore, in practice, a master and many slave control boards may be employed to substitute the central controller with powerful computing power, with the top-layer MPC incorporated in the master control board and the bottom-layer MPC algorithms implemented to m slave controllers. It demonstrates the excellent performance of the two-layer MPC technique for cell equalization real-time implementation.

Convergence Analysis: To analyze the convergence property of the designed two-layer MPC-based equalization approach, the Lyapunov stability theorem [14] is used. Because the ML equalizing current is identical for every cells in each module,

Table 8.1 Computational complexity comparison

Method	Optimization variables
Centralized MPC	$N(m+mn-1)$ coupled
Proposed two-layer MPC	$N(m-1)$ in the top layer
	$m \times Nn$ in the bottom layer

it has no effect on the SOC consistency of the cells in each module. Encouraged by this, we first demonstrate the convergence of the SOCs of the cells in each battery module, and then demonstrate that the SOC of the battery module can converge to the same value. The Lyapunov candidate functions are chosen as follows:

$$V_i(k) = || \bar{x}_i(k) - \bar{x}_i(k)||^2_{Q_i} \tag{8.26}$$

for $1 \le i \le m$. The global optimum of (8.25) is assumed to exist at each sampling step with optimal solution denoted as $U^*_{1,i}(k) = [u^{*T}_{1,i}(k), \ldots, u^{*T}_{1,i}(k+N-1)]^T$. It minimizes the cost function in (8.25) with the corresponding cost function denoted as $J_{b_i}(x_i(k), U^*_{1,i}(k))$. With a feasible control sequence at sampling step k as $U^f_{1,i}(k) = [0^T_n, u^{*T}_{1,i}(k) + u^{*T}_{1,i}(k+1), u^{*T}_{1,i}(k+2), \ldots, u^{*T}_{1,i}(k+N-1)]^T$, the cost function is denoted as $J_{b_i}(x_i(k), U^f_{1,i}(k))$. Therefore, the following inequality is to be satisfied

$$J_{b_i}(x_i(k), U^*_{1,i}(k)) \le J_{b_i}(x_i(k), U^f_{1,i}(k)) \tag{8.27}$$

From (8.24), it can be deduced that

$$x_i(k+1) - \bar{x}_i(k+1) = (I_{n,n} - \tfrac{1}{n}1_n 1^T_n)(x_i(k) + D_{1,i}u_{1,i}(k)) \tag{8.28}$$

where $I_{n,n}$ represents the identity matrix with dimensions of $n \times n$. Hence, for both cases of $[u^T_{1,i}(k), u^T_{1,i}(k+1)]^T = [u^{*T}_{1,i}(k), \ {}^{*T}_{1,i}(k+1)]^T$ and $[u^T_{1,i}(k), u^T_{1,i}(k+1)]^T = [0^T_n, u^{*T}_{1,i}(k) + u^{*T}_{1,i}(k+1)]^T$, the corresponding $x_i(k+2)$, the corresponding $x_i(k+2)$ is the same. From (8.27) and previous analysis, it yields the following:

$$\begin{aligned} &||x_i(k+1) - \bar{x}_i(k+1)||^2{}_{Q_i} + ||u^*_{1,i}(k)||^2{}_{R_i} + ||u^*_{1,i}(k+1)||^2_{R_i} \\ &\le ||x_i(k) - \bar{x}_i(k)||^2{}_{Q_i} + ||u^*_{1,i}(k) + u^*_{1,i}(k+1)||^2{}_{R_i}. \end{aligned} \tag{8.29}$$

By substituting (8.26) into (8.29), we can obtain the following:

$$V_i(k+1) - V_i(k) \le 2u^{*T}_{1,i}(k+1)R_i u^*_{1,i}(k). \tag{8.30}$$

If R_i is set as a zero matrix, the right side of the inequality in (8.30) is zero. Once the prediction horizon is set to $N = 1$ and a feasible control variable is chosen as $u^f_{1,i}(k) = 0_n$, the following inequality can be obtained, similar to the derivation process from (8.28) to (8.30).

$$V_i(k+1) - V_i(k) \leq -||u_{1,i}^*(k)||^2{}_{R_i} \tag{8.31}$$

with the right hand side is non-positive, and is equal to 0 only if $u_{1,i}^*(k) = 0_n$. Note that the optimal controlled CMC equalizing current is zero only when the cells' SOCs reach $||x_i(k) - \bar{x}_i(k)||^2 = 0$.

The closed-loop cell equalization system is stable based on (8.30) with $R_i = 0_{n,n}$ and asymptotically stable based on (8.31) with $N = 1$, according to the Lyapunov stability theorem. They are helpful in selecting control algorithm parameters. The prediction horizon in this work is set to $N = 1$, which reduces the computational difficulty of the designed two-layer MPC algorithm. Thereby, $x_i(k) \to \bar{x}_i(k)$ ($1 \leq i < m$), implying that the cells' SOCs can converge to their average value in each battery module. It can also be expressed as $x(k) \to [\bar{x}_1^T(k), \ldots, \bar{x}_m^T(k)]^T$.

Correspondingly, using the Lyapunov stability theorem, it can be demonstrated that $z(k) \to \bar{z}(k)$. Since $z(k) \otimes 1_n = [\bar{x}_1^T(k), \ldots, \bar{x}_m^T(k)]^T$ and $\bar{z}(k) \otimes 1_n = \bar{x}(k)$, it can be deduced that $x(k) \to \bar{x}(k)$. As a result, the closed-loop enhanced module-based CPC equalization system's convergence under the designed two-layer MPC technique is proved.

8.4 Results and Discussions

Thorough simulations in MATLAB/Simulink environment are carried out on a battery pack with 12 serially connected cells to validate the effectiveness of the proposed two-layer MPC-based equalization strategy, and the battery pack has been separated into three modules. The nominal capacity and voltage of the cells are set as 3.1 Ah and 3.7 V, respectively. The nominal voltages of the battery pack and each module are 44.4 V and 14.8 V. The weighing matrices are selected as $Q_c = 10^3 I_{m,m}$, $R_c = 0.1 I_{m-1,m-1}$, $Q_i = 10^3 I_{n,n}$, $R_i = 0.1 I_{n,n}$. Table 8.2 depicts the main parameters of the designed control algorithm. The FUDS [6] is used as the working current (as shown in Fig. 8.4a), with a magnitude of 1-C rate, i.e. 3.1 A. Once the RMS of the cells' SOC difference is less than $\epsilon = 0.25\%$, the equalization process is terminated to reduce the energy consumption of the equalizers. It should be noted that these parameters may also be assigned other appropriate values based on actual performance in practice.

Table 8.2 Parameter selections

Parameter	Value	Parameter	Value
I_{me_u}	1 A	$\alpha_{i,j}$	0.9
U_{2u}	1 A	$\beta_{i,j}$	0.9
U_{cm}	−3.1 A	β_{m_i}	0.9
U_{dm}	3.1 A	ϵ	0.25%
T	1 s	N	1

8.4.1 Equalization Results

The cells' initial SOC vector is chosen at random, as in case 1 in Table 8.3, with maximum and minimum SOCs of 86% and 62%, respectively. The RMS of the initial SOC difference between the cells is 6.821%. The RMS of the cells' SOC difference converges to 0.25% after 1090 s under the designed two-layer MPC-based equalization control strategy, as shown in Fig. 8.4b–c. The total currents of the cells are shown in Fig. 8.4d, demonstrating that the cell's current limitations can be adequately satisfied to ensure the cells' safety. The above results demonstrate that the propounded control technique can be effectively applied to the cell equalizing problem when the battery is powered by an external working current. The top-layer/bottom-layer MPC control inputs, i.e., the controlled ML/CMC equalizing currents, are shown in Fig. 8.5a–b.

Typically, the lowest SOC of the cells is defined as the SOC of the battery pack, because the battery pack must stop functioning once the SOC of one cell reaches zero. It can be observed that the SOC of the battery pack under the proposed equalization control strategy is 69.71% at 1090 s (the end of the equalization process), whereas the SOC of the corresponding battery pack without cell equalization is 57.85%. After equalization, the usable capacity can be as high as $11.86\% \times 3.1\ \text{Ah} = 0.37\ \text{Ah}$ which increases the battery pack capacity.

8.4.2 Comparison With the Centralized MPC

The results of the centralized MPC in (8.20) are also presented here as comparisons, with the cells' SOC responses shown in Fig. 8.6a. The total currents of the cells are depicted in Fig. 8.6b, illustrating that the current limitations can also be promised during the equalization process. The centralized MPC's controlled ML and CMC equalizing currents are shown in Fig. 8.6c– f, respectively. By making comparisons of Figs. 8.4b and 8.6a, it is possible to conclude that the consumed equalization time under the centralized MPC method (1078 s) is comparable to that under the designed two-layer MPC strategy (1090 s). It proves the outstanding performance of the designed two-layer MPC-based equalization technique, that can significantly

Table 8.3 Cases for different cells' initial SOC vectors

	Initial SOC vector [%]
Case 1	$[76, 74, 72, 66, 81, 77, 62, 78, 83, 75, 67, 86]^T$
Case 2	$[70, 90, 73, 69, 75, 68, 88, 76, 67, 79, 64, 71]^T$
Case 3	$[65, 90, 73, 74, 77, 76, 86, 68, 82, 62, 84, 80]^T$
Case 4	$[69, 72, 79, 67, 78, 87, 88, 74, 61, 64, 70, 83]^T$
Case 5	$[63, 80, 68, 78, 85, 76, 87, 82, 88, 69, 73, 62]^T$

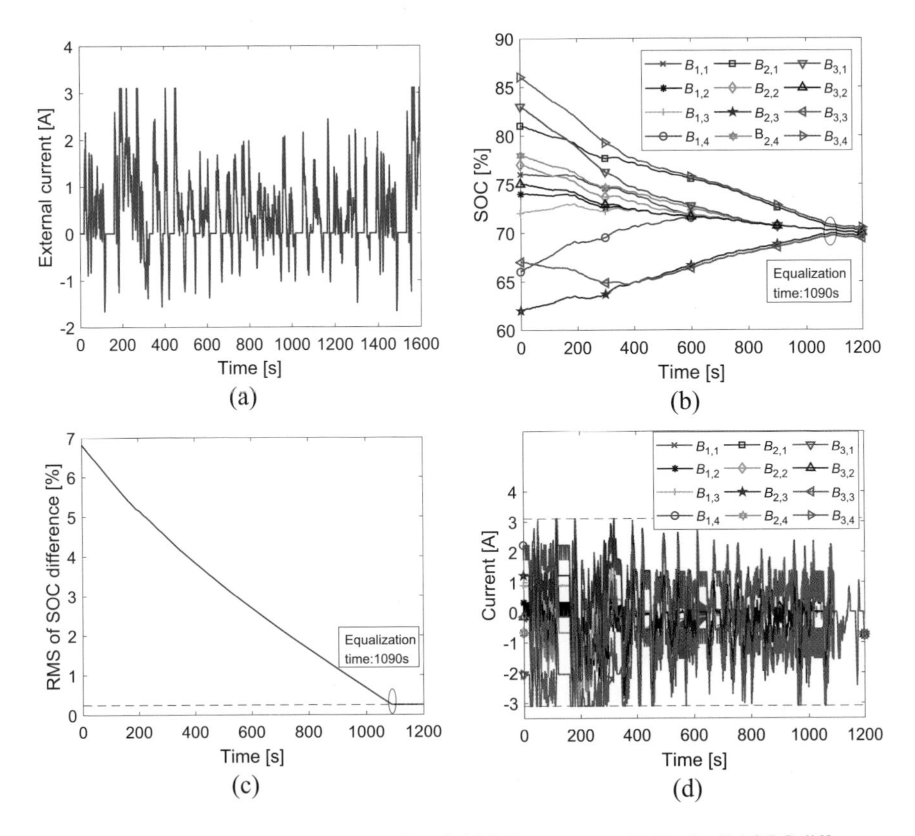

Fig. 8.4 **a** Battery pack's working current, **b** cells' SOC responses, **c** RMS of cells' SOC difference, **d** currents of the cells

decrease computational burden (as shown in Table 8.1) while slightly increasing equalization time, making real-time cell equalization implementation in practical applications more feasible.

8.4.3 Comparison With a Commercial CPC-Based Equalization Structure

For many commercial CPC-based active cell balancers, such as the LTC3300-1, one single balancer can only support the equalization for a limited number of serially connected cells. To achieve equalization for the battery pack consisting of a large number of serially-connected cells, the commonly used commercial CPC-based equalization systems usually divide the battery pack into several modules and adopt a interleaving connection of the module sides of the CMC equalizer to achieve balance between battery modules [1]. Referring to the commercial CPC-based equalization structure

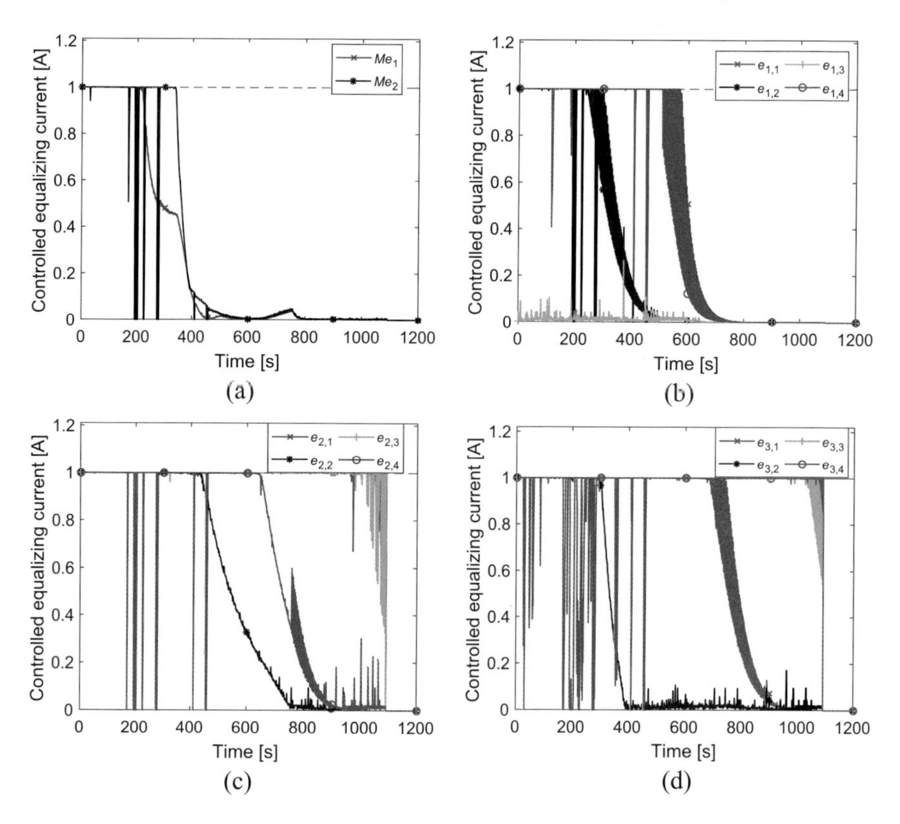

Fig. 8.5 Controlled equalizing currents of **a** ML equalizers, **b** CMC equalizer e_1, **c** CMC equalizer e_2, **d** CMC equalizer e_3 under the proposed two-layer MPC approach

with 3 battery modules in [1], 3 CMC equalizers are utilized, where the module sides of the first/second/ third CMC equalizers are connected with the cells 1 to 8/cells 5 to 12/cells 9 to 12, respectively. The cells' SOC responses and the RMS of the cells' SOC difference of the above battery pack under the bottom-layer MPC are illustrated in Fig. 8.7. The consumed equalization time is 3496 s, which is much higher than 1090 s as in Fig. 8.4. The comparison demonstrates the better performance of the developed improved module-based CPC equalization system.

8.4.4 Tests of Different Cells' Initial SOC Vectors

To further validate the effectiveness of the proposed two-layer MPC-based equalization strategy, other four tests are conducted on the battery pack with different cells' initial SOC vectors, where the initial SOC vectors are selected as shown in cases 2–5 in Table 8.3. The corresponding responses in terms of the cells' SOCs in the

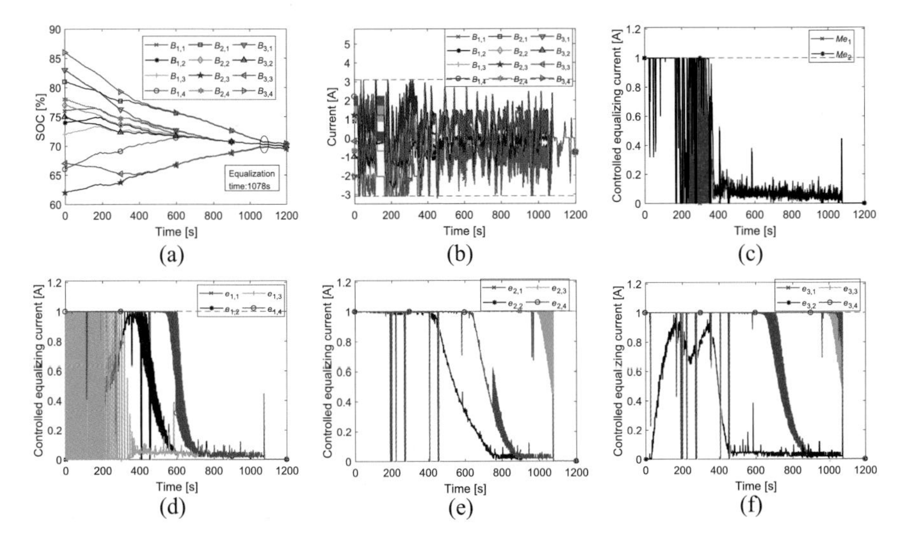

Fig. 8.6 **a** Cells' SOCs, **b** cells' currents, **c** controlled equalizing currents of ML equalizers, **d** controlled equalizing currents of CMC equalizer e_1, **e** controlled equalizing currents of CMC equalizer e_2, **f** controlled equalizing currents of CMC equalizer e_3 under the centralized MPC

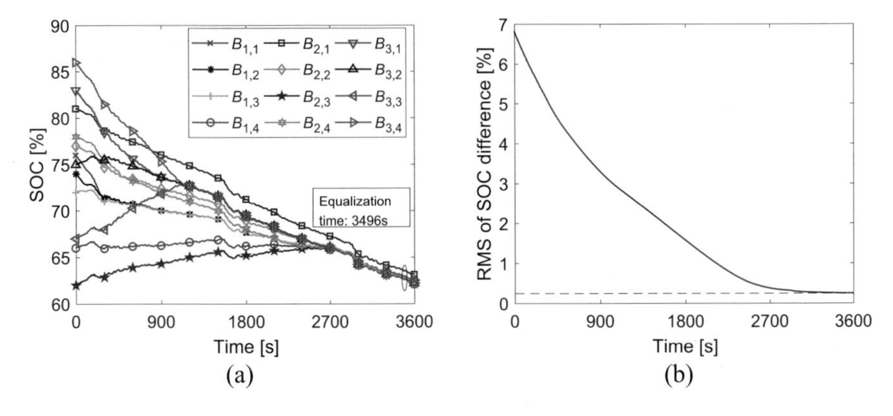

Fig. 8.7 **a** Cells' SOCs, **b** RMS of cells' SOC difference with the commercial CPC-based equalization structure

battery pack are summarized in Fig. 8.8, where their consumed equalization time is 1218 s, 1425 s, 1236 s, and 1483 s, respectively. From Fig. 8.8, it observes that the cells in the battery pack can be balanced for all cases, which further demonstrates the excellent performance of the designed improved equalization system and two-layer MPC approach.

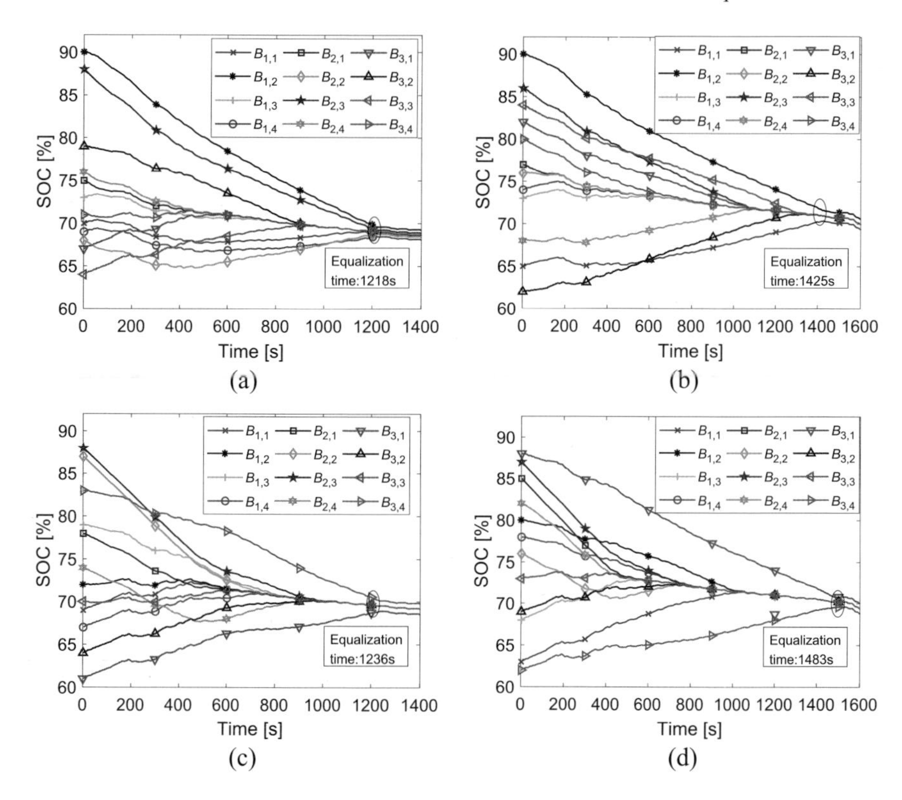

Fig. 8.8 SOC responses for the cells' initial SOC vector as in **a** Case 2, **b** Case 3, **c** Case 4, **d** Case 5 in Table 8.3

References

1. J. Drew, Active cell balancer extends run time and lifetime of large series-connected battery stacks. LT J. Analog. Innov. **23**(1), 3–7 (2013)
2. C. Pinto, J.V. Barreras, E. Schaltz, R.E. Araújo, Evaluation of advanced control for li-ion battery balancing systems using convex optimization. IEEE Trans. Sustain. Energy **7**(4), 1703–1717 (2016)
3. M. Einhorn, W. Guertlschmid, T. Blochberger, R. Kumpusch, R. Permann, F.V. Conte, C. Kral, J. Fleig, A current equalization method for serially connected battery cells using a single power converter for each cell. IEEE Trans. Veh. Technol. **60**(9), 4227–4237 (2011)
4. X. Tang, C. Zou, T. Wik, K. Yao, Y. Xia, Y. Wang, D. Yang, F. Gao, Run-to-run control for active balancing of lithium iron phosphate battery packs. IEEE Trans. Power Electron. **35**(2), 1499–1512 (2020)
5. S. Wang, L. Shang, Z. Li, H. Deng, J. Li, Online dynamic equalization adjustment of high-power lithium-ion battery packs based on the state of balance estimation. Appl. Energy **166**, 44–58 (2016)
6. H. He, R. Xiong, X. Zhang, F. Sun, J. Fan, State-of-charge estimation of the lithium-ion battery using an adaptive extended kalman filter based on an improved thevenin model. IEEE Trans. Veh. Technol. **60**(4), 1461–1469 (2011)

7. Q. Ouyang, J. Chen, J. Zheng, Y. Hong, Soc estimation-based quasi-sliding mode control for cell balancing in lithium-ion battery packs. IEEE Trans. Ind. Electron. **65**(4), 3427–3436 (2018)
8. Q. Ouyang, J. Chen, J. Zheng, H. Fang, Optimal cell-to-cell balancing topology design for serially connected lithium-ion battery packs. IEEE Trans. Sustain. Energy. **9**(1), 350–360 (2018)
9. Q. Ouyang, W. Han, C. Zou, G. Xu, Z. Wang, Cell balancing control for lithium-ion battery packs: a hierarchical optimal approach. IEEE Trans. Ind. Inform. **16**(8), 5065–5075 (2020)
10. C.F. Van Loan, The ubiquitous kronecker product. J. Comput. Appl. Math. **123**(1–2), 85–100 (2000)
11. C. Zou, C. Manzie, D. Nešic´, Model predictive control for lithium-ion battery optimal charging. IEEE/ASME. Trans. Mechatrons. **23**(2), 947–957 (2018)
12. J.B. Rawlings, D.Q. Mayne, *Model Predictive Control: Theory and Design* (Nob Hill Publishing, Madison, WI, USA, 2009)
13. S. Boyd, S.P. Boyd, L. Vandenberghe, Convex Optimization, (Cambridge university press, 2004)
14. H.K. Khalil, *Nonlinear Systems* (Prentice Hall, Englewood Cliffs, NJ, USA, 2002)

Chapter 9
Optimal Hierarchical Charging Equalization for Battery Packs

9.1 Charging System Model

Throughout this section, we consider a general charging scenario in which a battery pack can be charged using a variety of power sources, such as the a photovoltaic array, AC grid, and local energy storage. As shown in Fig. 9.1, a multi-module charger is designed for a n-modular serially connected battery pack made up of n modified isolated buck converters, each of which is used to charge one cell. The modified isolated buck converter-based charger has the advantages of being simple to implement, being small and inexpensive, and permitting for integrated infrastructure and modular design. Furthermore, the rapid advancement of integrated circuit technology has the potential to substantially reduce the cost of battery chargers, potentially allowing widespread use of the proposed multi-module charger.

9.1.1 Battery Pack Model

The system comprises a variety of battery models, which have been discussed in detail in [1]. Some of them include [2], an internal resistance equivalent circuit model that strikes a balance between accuracy and computational difficulty. It is utilized to describe each cell's dynamic behavior in the n-modular battery pack. It is worth noting that this model is generally appropriate for describing a battery operating within the normal temperature range. Therefore, the temperature's effects on battery variables have been overlooked here to simplify the analysis, design, and implementation of optimal charging control, although this propounded design can indeed be extended to include a thermal model. The battery model, as shown in Fig. 9.1, is made up of a voltage source and a serially connected internal resistor. The terminal voltage of the i-th cell can be written as:

$$V_{B_i} = V_{OC_i} + R_{0_i} I_{B_i} \tag{9.1}$$

J. Chen et al., *Equalization Control for Lithium-Ion Batteries*,
https://doi.org/10.1007/978-981-99-0220-0_9

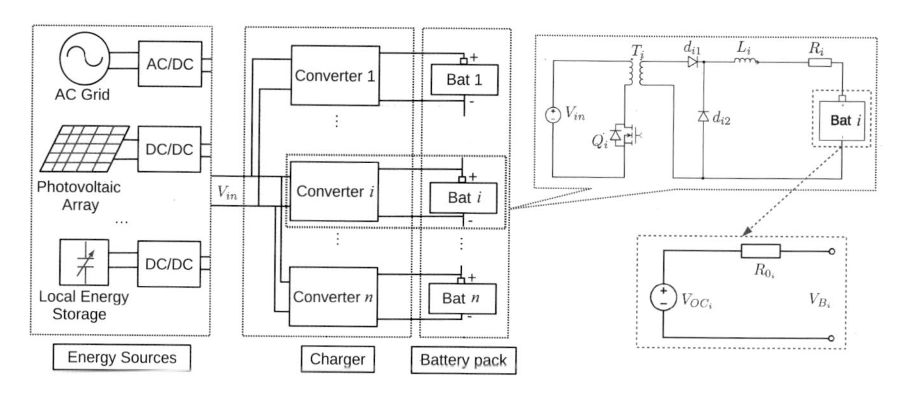

Fig. 9.1 Diagram of the multi-module charger for a serially connected lithium-ion battery pack

where V_{B_i} and V_{OC_i} represent the i-th cell's terminal voltage and OCV, respectively; R_{0_i} is the internal resistor; I_{B_i} is the current of the i-th cell, which is considered here positive if the cell is in the charging mode. The i-th cell's OCV and internal resistance are nonlinear functions of its SOC as $V_{OC_i} = f_i(SOC_i)$ and $R_{0_i} = h_i(SOC_i)$, where SOC_i denotes the i-th cell's SOC, and the dynamics of which is given by

$$\frac{\mathrm{d}SOC_i}{\mathrm{d}t} = \frac{\eta_i}{Q_i} I_{B_i} \tag{9.2}$$

where η_i represents the Coulombic efficiency, and Q_i is the i-th cell's capacity in Ampere-hour. The SOC of the i-th cell is assumed to be known throughout this chapter because it can be estimated with high accuracy by a large number of estimation algorithms, such as the extended Kalman filter [3] and the neural network based nonlinear observer [4]. The model of the n-modular battery pack can be expressed using (9.1)–(9.2) as follows:

$$\begin{aligned} \dot{x} &= B_1 v \\ y &= f(x) + h(x)v \end{aligned} \tag{9.3}$$

where the state vector is $x = [x_1, \ldots, x_n]^T \triangleq [SOC_1, \ldots, SOC_n]^T \in \mathbb{R}^n$; the input and output are $v = [v_1, \ldots, v_n]^T \triangleq [I_{B_1}, \ldots, I_{B_n}]^T \in \mathbb{R}^n$ and $y = [y_1, \ldots, y_n]^T \triangleq [V_{B_1}, \ldots, V_{B_n}]^T \in \mathbb{R}^n$; $f(\cdot) = [f_1(\cdot), \ldots, f_n(\cdot)]^T \in \mathbb{R}^n$; $B_1 = \mathrm{diag}\{\frac{\eta_1}{Q_1}, \ldots, \frac{\eta_n}{Q_n}\} \in \mathbb{R}^{n\times n}$ and $h(\cdot) = \mathrm{diag}\{h_1(\cdot), \ldots, h_n(\cdot)\} \in \mathbb{R}^{n\times n}$ with $\mathrm{diag}\{\cdot\}$ denoting the diagonal matrix.

9.1.2 Multi-module Charger Modeling

As shown in Fig. 9.1, a modified isolated buck converter design is proposed in this chapter, which comprised of a transformer T_i, an inductor L_i, a MOSFET Q_i, and two diodes d_{i1} and d_{i2}. Unlike traditional buck converters, this does not require a capacitor in the output port. When the continuous conduction mode is maintained [5], the i-th modified isolated buck converter mathematical model is formulated as follows:

$$\frac{\mathrm{d}I_{B_i}}{\mathrm{d}t} = \frac{D_i}{L_i}V_{in} - \frac{V_{B_i}}{L_i} - \frac{I_{B_i}R_i}{L_i} \tag{9.4}$$

where V_{in} represent the rectified DC input voltage, R_i is a current sense resistor, V_{B_i} and I_{B_i} represent the i-th cell's terminal voltage and charging current, and D_i $(0 \le D_i \le 1)$ is the duty cycle of the PWM signal applied to MOSFET Q_i. In reference of (9.4), the multi-module charger model can be represented as follows:

$$\dot{v} = -Av + C_1u - C_2y \tag{9.5}$$

with $A = \text{diag}\{\frac{R_1}{L_1}, \ldots, \frac{R_n}{L_n}\} \in \mathbb{R}^{n\times n}$, $C_1 = \text{diag}\{\frac{V_{in}}{L_1}, \ldots, \frac{V_{in}}{L_n}\} \in \mathbb{R}^{n\times n}$, $C_2 = \text{diag}\{\frac{1}{L_1}, \ldots, \frac{1}{L_n}\} \in \mathbb{R}^{n\times n}$, where $u = [u_1, \ldots, u_n]^T \triangleq [D_1, \ldots, D_n]^T \in \mathbb{R}^n$ is the control variable.

Remark 9.1 To avoid overcharging, conventional chargers that deliver the same charging current to all cells in the battery pack must terminate the charging process when one of the cells reaches the upper SOC threshold, even if the other cells are not fully charged. This will result in much more conservative use of the battery pack. Each cell in our proposed multi-module charger is charged independently by a modified isolated buck converter. This design will provide two advantages that differentiate it from the others available in the literature. The first advantage is that the charging control can be imposed at the cell level, giving us greater control over the charging process. The second advantage of being able to control individual cells is that we can delicately regulate the charging process for each cell in order to balance the SOC toward the same level.

9.1.3 Charging System Modeling

The overall charging system model consists of the model of the battery pack and the model of the multi-module charger (9.5), where the charger system's output is the input of the battery pack system. The duty cycle vector u in (9.5) is controlled to regulate the charging current vector v to make the cells' SOC vector reach its desired value. Because the modified isolated buck converter-based charger has a much faster transient response than the battery pack's dynamics, the control strategy can be divided into two layers: the top layer for generating the optimal charging current profile and the bottom layer for controlling the charger to provide the desired charging currents. The development of such a hierarchical control technique will be presented next.

9.2 Hierarchical Control for the Charging Equalization System

In this section, the objective functions and constraints for the charging process's will be formulated. Thereafter, a hierarchical control technique is proposed to ensure that the charging system can meet the defined objectives and constraints. The overall hierarchical control approach consists of two layers, as depicted in Fig. 9.2. The top-layer control is responsible for scheduling the optimal charging currents while taking into account cell equalization, user demand, charging process constraints, and temperature buildup. The bottom-layer control is responsible for regulating the charging currents provided by the designed multi-module charger in order to track the charging currents specified by the top layer.

The battery pack model (9.3) is discretized to better schedule the optimal charging currents by holding the charging currents constant during each scheduling sampling interval. The discrete-time model is represented by

$$
\begin{aligned}
x(k+1) &= x(k) + Bv(k) \\
y(k) &= f(x(k)) + h(x(k))v(k)
\end{aligned}
\tag{9.6}
$$

where $B = \text{diag}\{\frac{\eta_1 T}{Q_1}, \ldots, \frac{\eta_n T}{Q_n}\} \in \mathbb{R}^{n \times n}$ with the scheduling sampling period T.

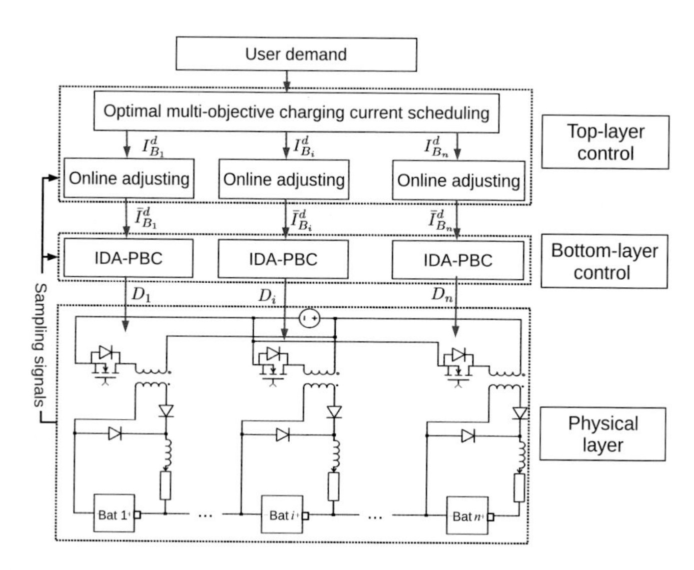

Fig. 9.2 Block diagram of the overall battery charging process

9.2.1 Charging Equalization Objectives

Charging task: For a battery pack with the initial cells' SOC vector $x(0) = x_0$, the user can specify the target SOC and charging duration as follows:

$$x_s(N) = \Gamma_{set} 1_n \text{ with } T_{set} = NT \tag{9.7}$$

where Γ_{set} and T_{set} are the desired SOC and the demanded charging duration of the battery pack, respectively; 1_n represents a column vector with n ones and N is the sampling step number. It is intended to minimize the difference between $x(N)$ and $x_s(N)$ in order to meet this charging objective, and the associated cost function can be expressed as:

$$J_x = \frac{1}{2}(x(N) - x_s(N))^T (x(N) - x_s(N)). \tag{9.8}$$

In [6], a hard constraint of the terminal state vector $x(N) = x_s(N)$ is used to make the cells' SOCs at the end of the charging process equal to their target value. However, the user settings may not be accomplished in practice; for example, even if we persistently charge the battery with the maximum allowed charging current, the battery pack cannot be fully charged in a very short charging duration of the user demand. In these cases, using the hard constraint in our charging current scheduling algorithm is not appropriate because it might result in an infeasible problem with the solution set being empty under all charging constraints. Such nonsense cases can be effectively avoided by using the cost function (9.8) instead of the hard constraint $x(N) = x_s(N)$, so that a feasible solution always exists to minimize J_x and drive $x(N)$ as close to the desired $x_s(N)$ as possible while satisfying all charging constraints.

Cell equalization: The energy imbalance among cells in a battery pack exists due to imperfect battery manufacturing technology and spatially uneven temperature distribution. The lowest/highest SOC of the cells limits the overall battery pack's usable/rechargeable capacity. As a result, by bringing the SOC of each cell to the same level, the effective capacity of the battery pack can be significantly increased. The cell equalization problem can indeed be reframed as minimizing the cells' SOC difference $||x(k) - \text{bar}x(k)||$ during the charging process, where $||\text{cdot}||$ represents the 2-norm and $\text{bar}x(k) = \text{frac } 1n1_n1_n{}^T x(k)$ represents the battery pack's average SOC vector. As a result, the cost function for cell equalization can be written as:

$$J_e = \frac{1}{2N} \sum_{k=1}^{N} x^T(k) D^T D x(k) \tag{9.9}$$

with $D = I_n - \frac{1}{n} 1_n 1_n^T$, where I_n represents the identity matrix with dimensions of $n \times n$. In (9.9), $\frac{1}{N}$ is included to make the order of magnitude of J_e comparable to J_x in (9.8).

Temperature effects: Based on [7], the temperature dynamics of the i-th ($1 \leq i \leq n$) cell can be modeled as:

$$T_{e_i}(k+1) = T_{e_i}(k) - b_{i1}(T_{e_i}(k) - T_a) + b_{i2}v_i^2(k) + \sum_{j=1}^{n} r_{ij}(T_{e_j}(k) - T_{e_i}(k)) \tag{9.10}$$

where $T_{e_i}(k)$ is the i-th cell's lumped temperature, T_a denotes the ambient temperature, r_{ij} represents the coefficient of the heat transfer from the j-th cell to the i-th cell in the battery pack, b_{i1} and b_{i2} are coefficients relevant to heat transfer and Joule heating, respectively. The temperature rise over a sampling period can be defined as $\Delta T_e(k) = T_e(k+1) - T_e(k)$, where $T_e(k) = [\, T_{e_1}(k), \ldots, T_{e_n}(k)]^T \in \mathbb{R}^n$. Based on (9.10), it can be obtained that

$$\Delta T_e(k) = A_e T_e(k) + B_{e1} T_a + B_{e2} \underline{v}(k) \tag{9.11}$$

where $A_e = [-b_{11} + r_{11} - \sum_{j=1}^{n} r_{1j}, r_{12}, \ldots r_{1n}; \quad r_{21}, -b_{21} + r_{22} - \sum_{j=1}^{n} r_{2j}, \ldots, r_{2n}; \ \ldots; \ r_{n1}, r_{n2}, \ldots, -b_{n1} + r_{nn} - \sum_{j=1}^{n} r_{nj}] \in \mathbb{R}^{n \times n}$, $B_{e1} = [b_{11}, b_{21}, \ldots, b_{n1}]^T \in \mathbb{R}^n$, $B_{e2} = \text{diag}\{b_{12}, b_{22}, \ldots, b_{n2}\} \in \mathbb{R}^{n \times n}$, and $\underline{v}(k) = [v_1^2(k), \ldots, v_n^2(k)]^T \in \mathbb{R}^n$. The temperature rises of the cells can be controlled by using (9.11) to build a cost function. However, because r_{ij}, b_{i1}, and b_{i2} in (9.11) are difficult to identify accurately in practice, obtaining the precise temperature rise is difficult. Furthermore, using a model like that in (9.11) will substantially elevate the computational cost while tackling the charging optimization problem formulated later. However, according to (9.11), the temperature rises of the cells are proportional to the square of their charging currents. Charging currents should not be too high to keep the temperature of the cells within the normal operating range. As a result, in order to simplify the thermal rise management problem, we seek to constrain the magnitude of the charging current squared rather than the temperature rise directly. This results in the following cost function:

$$J_v = \frac{1}{2N} \sum_{k=0}^{N-1} v^T(k) v(k). \tag{9.12}$$

In most cases, when the cell currents and terminal voltages are strictly less than the maximum allowed values, the battery pack's temperature will not exceed its limit. As a result, we do not need the hard temperature restriction in the charging current design algorithm and instead use (9.12) to ensure that the temperature rise of the battery pack does not become too high. When the temperatures of the cells exceed their maximum, the charging process is stopped for thermal safety.

Multi-objective formulation: The user demand, cell equalization, and thermal effect should all be considered when determining the optimal charging current scheduling issue. To balance them, the quadratic multi-objective cost function (9.8), (9.9), and (9.12) are combined as follows:

$$J(x(k), v(k)) = \gamma_1 J_x + \gamma_2 J_e + \gamma_3 J_v \tag{9.13}$$

where $\gamma_1 \geq 0$, $\gamma_2 \geq 0$, and $\gamma_3 \geq 0$ are the trade-off weights.

9.2.2 Charging Constraints

To ensure the battery pack system's stability and longevity, the following three constraints must be met during the charging process: cell SOC limitations, charging current restrictions, and terminal voltage constraints.

SOC limitations: To avoid overcharging the battery pack, the SOCs of the cells should be below the upper limit x_u.

$$\chi = \{x(k) \in \mathbb{R}^n | x(k) \leq x_u\}. \tag{9.14}$$

Charging current restrictions: Because excessive charging currents, defined as charging currents greater than the maximum allowed values, are harmful to batteries, the charging currents of the cells should be kept within a certain suitable range ν such that

$$\nu = \{v(k) \in \mathbb{R}^n | 0_n \leq v(k) \leq v_M\} \tag{9.15}$$

where $v_M \in \mathbb{R}^n$ denotes the cells' maximum allowed charging current vector, and 0_n denotes the vector with n zeros.

Terminal voltage constraints: The terminal voltages of the cells should not exceed the upper voltage limit. It should satisfy (9.3) for the battery pack as follows:

$$f(x(k)) + h(x(k))v(k) \leq y_M \tag{9.16}$$

where $y_M \triangleq V_{max} 1_n \in \mathbb{R}^n$ with V_{max} the cells' maximum allowed terminal voltage.

9.2.3 Top-Layer Control: Optimal Charging Current Scheduling

With the multi-objective (9.13) and constraints (9.14)–(9.16) in the charging process taken into account, the charging current scheduling problem can be transformed to a constrained optimization problem as follows:

$$\begin{aligned} \underset{v(0),\ldots,v(N-1)}{\text{minimize}} \quad & J(x(k), v(k)) \\ \text{subject to} \quad & x(k+1) = x(k) + Bv(k), \quad x(0) = x_0 \\ & f(x(k)) + h(x(k))v(k) \leq y_M \\ & x(k) \leq x_u, \quad v(k) \in \nu. \end{aligned} \tag{9.17}$$

According to (9.17), it yields $x(k) = x_0 + \sum_{j=0}^{k-1} Bv(j)$, which can be rewritten as $x(k) = x_0 + BH_kU$ by defining $U \triangleq [v^T(0), \ldots v^T(N-1)]^T \in \mathbb{R}^{nN}$ and $H_k \triangleq [\Upsilon_k, \Theta_k] \in \mathbb{R}^{n \times nN}$ with $\Upsilon_k = [I_n, \ldots, I_n] \in \mathbb{R}^{n \times kn}$ and $\Theta_k = [0_{n \times n}, \ldots, 0_{n \times n}] \in$

$\mathbb{R}^{n\times(N-k)n}$. The constrained optimization problem (9.17) can then be rewritten as follows:

$$\begin{aligned} \underset{U}{\text{minimize}} \quad & J_1(U){+}\tau \\ \text{subject to} \quad & F(U)+G(U)U \leq Y_M \\ & MU \leq X_C, \quad \Phi U \leq U_M \end{aligned} \tag{9.18}$$

with the auxiliary variables

$$J_1(U) = x_0^T \Psi_1 U{-}x_s^T \Psi_2 U{+}U^T \Psi_3 U$$

$$\tau = \frac{\gamma_1}{2}(x_0^T x_0{-}2x_0^T x_s(N){+}x_s^T(N)x_s(N)){+}\frac{\gamma_2}{2}x_0^T D^T D x_0$$

$$\Psi_1 = \gamma_1 B H_N{+}\frac{\gamma_2}{N}\sum_{j=1}^{N} D^T D B H_j, \Psi_2 = \gamma_1 B H_N$$

$$\Psi_3 = \frac{\gamma_1}{2}H_N^T B^T B H_N{+}\frac{\gamma_2}{2N}\sum_{j=1}^{N} H_j^T B^T D^T D B H_j{+}\frac{\gamma_3}{2N}I_{nN}$$

where U is the optimization variable; $Y_M \in \mathbb{R}^{nN}$, $X_C \in \mathbb{R}^{nN}$, $F(U) \in \mathbb{R}^{nN}$, $G(U) \in \mathbb{R}^{nN\times nN}$, $M \in \mathbb{R}^{nN\times nN}$, $\Phi \in \mathbb{R}^{2nN\times nN}$, and $U_M \in \mathbb{R}^{2nN}$ are defined as:

$$Y_M = [y_M^T, \ldots, y_M^T]^T, X_C = [(x_u - x_0)^T, \ldots, (x_u - x_0)^T]^T$$

$$F(U) = [f^T(x_0{+}DH_1U), \ldots, f^T(x_0{+}DH_NU)]^T$$

$$G(U) = \text{diag}\{h(x_0{+}DH_1U), \ldots, h(x_0{+}DH_NU)\}$$

$$M = [(BH_1)^T, \ldots (BH_N)^T]^T, \Phi = [I_{nN}, -I_{nN}]^T$$

$$U_M = [v_M^T, \ldots, v_M^T, 0_n^T, \ldots, 0_n^T]^T.$$

Because tau is a constant, it can be removed to solve the optimization problem (9.18). The constrained optimization (9.18) can be transformed as follows by introducing a convex barrier function [8] to replace the inequality constraints:

$$\underset{U}{\text{minimize}}\ J_1(U){-}\frac{1}{\mu_1}\sum_{j=1}^{4nN}\log({-}g_j(U)) \tag{9.19}$$

where $-\sum_{j=1}^{4nN}\log(-g_j(U))$ represents the logarithmic barrier function, μ_1 is a positive parameter, and $g_j(U)$ is the j-th element of the vector $g(U) \in \mathbb{R}^{4nN}$ given as:

$$g(U) = \begin{bmatrix} F(U)+G(U)U-Y_M \\ MU-X_C \\ \Phi U-U_M \end{bmatrix}. \tag{9.20}$$

The optimization problem (9.19) can be solved using an interior point-based algorithm in [8], and the sequence of optimal charging currents $I_{B_i}^d$ $(1 \le i \le n)$ can be obtained by off-line computation.

Variable scheduling sampling period: When used in practice, the scheduling sampling period is critical to the success of the charging control approach. A long sampling period T in scheduling implies a large model discretization error of (9.6). As a result, when the cells are charged close to fully charged, their actual states may implicitly exceed the limitations. However, if the scheduling sampling period is too short, the computational cost will be too high for practical implementation. A variable scheduling sampling period is used to strike a balance between performance and calculation time.

$$T = \begin{cases} T_1, \; x(k) \le x_T \\ T_2, \; x(k) > x_T \end{cases} \tag{9.21}$$

where $x_T = \Gamma_T 1_n$ with Γ_T is a preset SOC that is close to the fully-charged state, T_1 and T_2 are scheduling sampling periods that can be designed by taking into account both the practical accuracy requirement of the discrete-time battery model and the computational complexity of the charging current scheduling algorithm. As depicted in (9.21), a large T_1 is selected to reduce the computational time when the SOCs of the cells are less than x_T, and the scheduling sampling period is changed to a small value T_2 to avoid large model discretization error when the SOCs of the cells increase to be greater than x_T. Algorithm 1 describes the detailed method for implementing the variable scheduling sampling period.

Algorithm 1:

- (1) Set the user demand Γ_{set}, T_{set}, the initial cells' SOC vector x_0, the preset SOC vector x_T, and the scheduling sampling period $T = T_1$. The step number is $N_1 = \frac{T_{set}}{T_1}$.
- (2) Run the interior point algorithm and get the scheduled cells' charging current vector $v(k)$ $(k = 0, \ldots, N_1 - 1)$ and the updated SOC vector $x(k)$ $(k = 1, \ldots, N_1)$.
- (3) If $\Gamma_{set} 1_n \le x_T$, stop and output $v(k)$ $(k = 0, \ldots, N_1 - 1)$. Otherwise, increase k from 1 to N_1 to find a k_1 that satisfies $k_1 = \max\{k : x(k) \le x_T\}$. Record $v(k)$ $(k = 0, \ldots, k_1 - 1)$ and set $x(k_1)$ as the new initial cells' SOC vector.
- (4) Change the scheduling sampling period $T = T_2$, and the step number is $N_2 = \frac{T_{set} - k_1 T_1}{T_2}$. Run the interior point algorithm again and get the designed cells' charging current vector $v(k)$ ($k = k_1, \ldots, k_1 + N_2 - 1$).
- (5) Terminate and output $v(k)$ ($k = 0, \ldots, k_1 - 1, k_1, \ldots, k_1 + N_2 - 1$) by combining the results from Step (3) and Step (4).

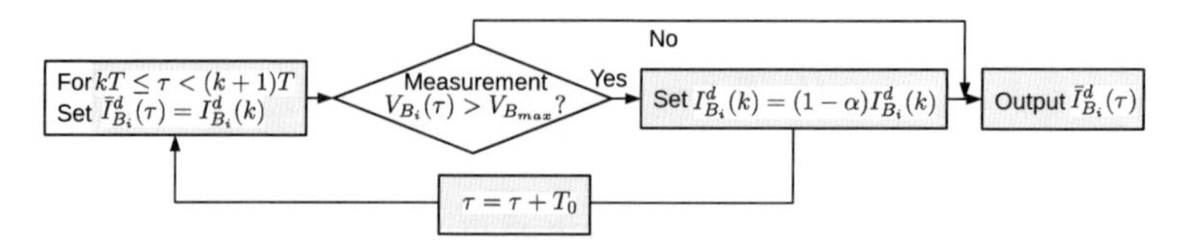

Fig. 9.3 Flow chart of the on-line charging current adjustment strategy

Online adjustment of desired charging currents: The i-th ($1 \le i \le n$) cell's terminal voltage V_{B_i} may exceed its upper bound with the scheduled optimal charging currents $I^d_{B_i}$ due to the existence of bias and disturbance between the model in the charging current scheduling algorithm and the actual characteristics of the battery pack in practice. As shown in (9.16), the terminal voltage increases with charging current, implying that a high charging current results in a high terminal voltage. As a result, once the i-th cell's measured terminal voltage V_{B_i} exceeds its maximum allowed value $V_{B_{max}}$, the charging current should be reduced to cause V_{B_i} to fall. Based on this concept, an on-line charging current adjusting strategy is proposed that reduces the actual desired charging current by a specific rate α ($0 < \alpha < 1$) until the i-th cell's measured terminal voltage V_{B_i} meets the constraints (9.16). Figure 9.3 depicts the technique, where $\bar{I}^d_{B_i}$ is the desired charging current after adjustment for the i-th cell and is set as the bottom-layer controller's tracking value, and T_0 is the control sampling period. It should be noted that a higher adjusting rate α results in faster tuning speed, which may result in less overvoltage duration but may result in an excessive reduction of $\bar{I}^d_{B_i}$, which consists the charging process performance. Through experimental validation, a suitable α with a fast tuning speed to ensure the cells' terminal voltage constraints without sacrificing much performance can be chosen. Because the overvoltage phenomenon occurs only when the cells are nearing full charge, the actual cells' SOC trajectories are still close to the scheduled ones.

9.2.4 Bottom-Layer Control: Charging Current Tracking

The bottom-layer controller's goal is to make the actual charging currents through the multi-module charger track the top layer's desired values $\bar{I}^d_{B_i}$ ($1 \le i \le n$). From an energy standpoint [5], we can define the charger's Hamiltonian function to denote the total energy stored in the system as:

$$H(v) = \tfrac{1}{2} v^T C v \tag{9.22}$$

with $C = \mathrm{diag}\{L_1, \ldots, L_n\} \in \mathbb{R}^{n \times n}$. Based on (9.5) and (9.22), the model of the charger can be rewritten as:

$$\dot{v} = -A_1 \tfrac{\partial H(v)}{\partial v} + C_1 u - C_2 y \tag{9.23}$$

with $A_1 = \text{diag}\{\frac{R_1}{L_1^2}, \ldots, \frac{R_n}{L_n^2}\} \in \mathbb{R}^{n\times n}$, where y is considered as a measurable disturbance vector here. To ensure the charging currents v_i $(1 \le i \le n)$ tracking their desired values $\bar{I}_{B_i}^d$, an IDA-PBC is proposed for the system (9.23), which assigns a desired energy function that has a minimum at the desired equilibrium point by modifying the control input [5]. Here, we define the desired total energy function as:

$$H_d(v) = \frac{1}{2}(v - v^d)^T C(v - v^d) \tag{9.24}$$

where $v^d = [\bar{I}_{B_1}^d, \ldots, \bar{I}_{B_n}^d]^T \in \mathbb{R}^n$. Because the desired charging current is a slow-changing signal in comparison to the fast dynamics of the converter, the derivative of the desired charging current can be assumed to be zero and thus $\dot{v}^d = 0_n$. From (9.24), it shows that $H_d(v) \ge 0$ and $H_d(v) = 0$ if and only if $v = v^d$ since C is positive definite. Assuming that a control input $u = \beta(v, v^d)$ can be found that makes the closed-loop system of (9.23) satisfy

$$\dot{v} = -R_d \frac{\partial H_d(v)}{\partial v} \tag{9.25}$$

where $R_d = \text{diag}\{R_{d_1}, \ldots, R_{d_n}\} \in \mathbb{R}^{n\times n}$ is a designed positive definite matrix. According to (9.23) and (9.25), it yields

$$-R_d \frac{\partial H_d(v)}{\partial v} = -A_1 \frac{\partial H(v)}{\partial v} + C_1\beta(v, v^d) - C_2 y. \tag{9.26}$$

From (9.26), the control input can be deducted as:

$$u = \beta(v, v^d) = C_1^{-1}((A_1 - R_d)Cv + R_d C v^d + C_2 y) \tag{9.27}$$

where C_1^{-1} denotes the inverse matrix of C_1.

Stability proof: A Lyapunov candidate function is selected as follows:

$$V = H_d(v) \tag{9.28}$$

where $H_d(v)$ is seen in (9.24). According to (9.22)–(9.27), the derivative of (9.28) satisfies

$$\dot{V} = \frac{\partial H_d(v)}{\partial v}\dot{v} \le -P\tilde{v}^2 \tag{9.29}$$

where $P = CR_dC$, $\tilde{v} = (v - v^d)$. Since P is positive definite, $\dot{V} \le 0$ and $\dot{V} = 0$ if and only if $v = v^d$. According to the LaSalle's invariance principle [9], $v \to v^d$. Hence, the stability proof of the closed-loop bottom-layer charging current tracking system is achieved.

Since the modified buck converters are actually independently controlled, the controller (9.27) can be rewritten as n distributed IDA-PBCs:

$$u_i = \frac{(R_i - R_{d_i} L_i^2) v_i + R_{d_i} L_i^2 v_i^d + V_{B_i}}{V_{in}} \tag{9.30}$$

for $1 \leq i \leq n$, where R_{d_i} is chosen to make the duty cycle u_i within the set [0, 1].

Remark 9.2 The actual charging currents via the charger are assured to asymptotically track their desired values as scheduled by the top layer, based on the above stability proof. Because the top layer can ensure the charging constraints of the battery pack system with the scheduled charging currents, the proposed hierarchical approach can achieve stable battery charging with the actual charging currents.

9.3 Simulation and Experimental Results

The proposed battery charger and the optimal charging current scheduling approach are validated in this section through extensive simulation and experiments. The experimental bench, as shown in Fig. 9.4a, is primarily comprised of a battery pack consisting of three serially connected PowerFocus 18650 lithium-ion batteries, a set of dSPACE, three OMEGA temperature sensors, and an NI GPIC Single-Board 9683.

With many charging experiments, the cells capacity are found as $Q_1 = 2.341$ Ah, $Q_2 = 2.387$ Ah, and $Q_3 = 2.379$ Ah, respectively. The relationships between the cells' OCVs and SOCs are depicted in Fig. 9.5a, and the internal resistances of the cells are depicted in Fig. 9.5b. Internal resistances of cells are nearly constant within the SOC range [5%, 95%], but sharply increase when the cells' SOCs are close to fully charged states [95%, 100%]. This implies that the terminal voltages of the cells may easily exceed their limits with SOCs greater than 95%. As a consequence, the preset SOC in the variable scheduling sampling method is $\Gamma_T = 95\%$. The sampling times

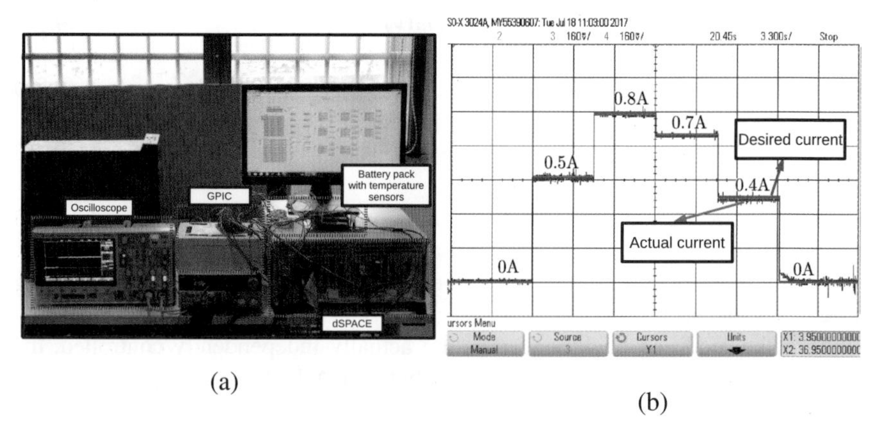

Fig. 9.4 **a** Experimental test bench, **b** desired and actual current responses of the modified buck converter with the IDA-PBC

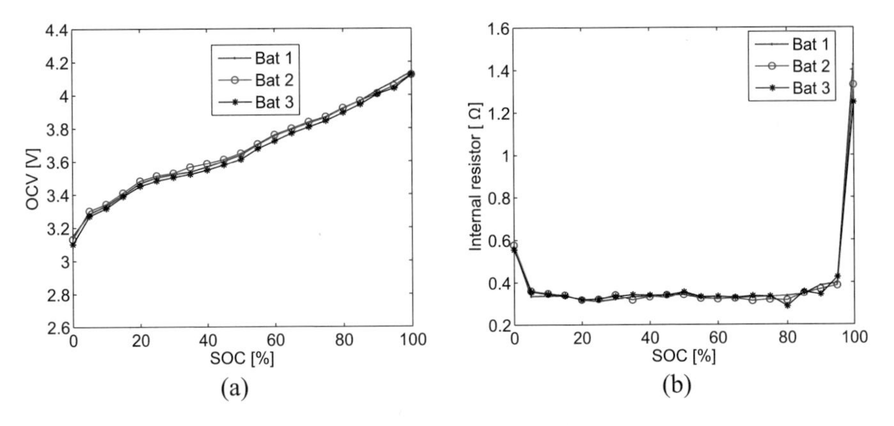

Fig. 9.5 **a** Relationship between the cells' OCVs and SOCs, **b** mapping from the SOCs to the internal resistors of the cells

are set to $T_1 = 600$ s and $T_2 = 60$ s. The cells' SOCs have an upper bound of $x_u = 100\%$, and their maximum allowed terminal voltage is set to $V_{max} = 4.2$ V. The GPIC is used to build the on-line adjustment of desired charging currents. If the measured terminal voltage of the cell is greater than 4.2 V, the desired charging current is reduced by 5% with the control sampling period $T_0 = 0.5$ s until the measured terminal voltage of the cell is pushed below the constraints (9.16). It should be noted that other appropriate parameters can be designated depending on the requirements and performance in practice.

The modified isolated buck converter based multi-module charger is modeled in the dSPACE using real-time simulation with inductor $L_i = 0.01$ H $(1 \leq i \leq 3)$, non-ideal transformer T_i, MOSFET Q_i with internal diode resistance 0.1 Ω and FET resistance 0.01 Ω, diodes d_{i1} and d_{i2} with snubber resistances 500 Ω. The DC input voltage is $V_{in} = 24$ V, and the frequency of the PWM signal applied on the MOSFETs is set as 5 kHz. The current sense resistor is chosen as $R_i = 1$ Ω. First, a preliminary experiment is performed to validate the performance of the modified isolated buck converter. The desired and actual charging current responses of the modified buck converter with the IDA-PBC are shown in Fig. 9.4b, in which the actual current is obtained by measuring the voltage across the current sense resistor with a Keysight oscilloscope. The experimental results show that the designed IDA-PBC can track the desired charging current through the converter at a satisfactory level.

9.3.1 Simulation Results

To validate the effectiveness of the propounded optimal charging current scheduling algorithm, MATLAB simulations for a 10-modular serially connected battery are performed. The capacities of the cells are chosen at random, ranging from 1.90 Ah to 2.10 Ah with $[Q_1, \dots, Q_{10}] = [2.07,1.91,1.93, 1.96,1.97,2.09,2.06, 1.99,1.95,2.10$ Ah]. The cells' initial SOCs are randomly given as $[x_1(0), \dots, x_{10}(0)] = [12, 20, 18, 19, 16, 14, 17, 10, 15, 11\%]$. The initial cells' SOC difference is $||x(0) - \bar{x}(0)|| = 10.28\%$. The maximum allowed charging current for the cells is set to 3 C. Without sacrificing generality, the OCVs and internal resistances of the simulation's first, fourth, seventh, and tenth/second, fifth, and eighth/third, sixth, and ninth cells are selected to be the same as the model parameters of the actual battery cell 1/2/3 in the experiment (fitted by interpolation) shown in Fig. 9.5. The ambient temperature is set as 25 °C and the temperature influence coefficients b_{i1} and b_{i2} $(1 \leq i \leq 10)$ are selected as the same in [7]. To explain and analyze the performance of the designed charging technique, the desired SOC of the battery pack has been selected as $\Gamma_{set} = 100\%$, and the charging durations are set as $T_{set} = 60$ min, $T_{set} = 120$ min, and $T_{set} = 180$ min, respectively. The weights of the cost function (9.13) in the optimal charging current scheduling algorithm are selected as $\gamma_1 = 1000$, $\gamma_2 = 5$, and $\gamma_3 = 0.01$.

The simulation results in terms of cell SOCs, charging currents, terminal voltages, and temperatures are shown in Fig. 9.6. The designed optimal charging currents in our proposed charging strategy can be adjusted with different user settings, as shown in Fig. 9.6b, f, and j. Large charging currents are designed to charge the cells' SOCs near to the desired SOC with high cell temperatures (maximum temperature of $29.01 rmoC$ in the charging process) for a short and tight charging duration of 60 min. Small charging currents, on the other hand, are utilized for a long and adequate charging duration of 180 min, which meets the charging requirement while resulting in comparatively low temperatures (maximum temperature of $26.81\, rmoC$). The terminal voltage responses of the cells are depicted in Fig. 9.6c, g, and k, that can satisfy the constraints (9.16) in accordance with the designed charging currents. Table 9.1 for different user settings calculates the cells' average SOC and SOC difference at the end of the charging process, as well as the maximum temperature during the charging process. With the designed optimal charging currents, the cells' average SOC can be charged to close to the desired SOC. The effectiveness of the optimal charging current scheduling strategy designed is also illustrated on cell equalization.

High C-rate current charging: Because the cells in the simulation have high internal resistances, the charging currents are limited to less than 1 C-rate to avoid the overvoltage phenomenon shown in Fig. 9.6, despite the fact that the maximum allowed charging current is set at 3 C-rate. To illustrate how the designed charging strategy performs at higher C-rates, the internal resistances in the simulation are set to 10% of their original values. The desired SOC and charging time are set to $\Gamma_{set} = 100\%$ and $T_{set} = 20$ min, respectively. The SOCs, charging currents, terminal voltages, and temperatures of the cells are depicted in Fig. 9.7. The cells' largest charging currents can reach 3 C-rate, and their average SOC rises to 94.14% after $T_{set} = 20$ min charging. During the charging cycle, the maximum temperature of the cells is around 38.8 °C, which meets the actual thermal constraints. The results

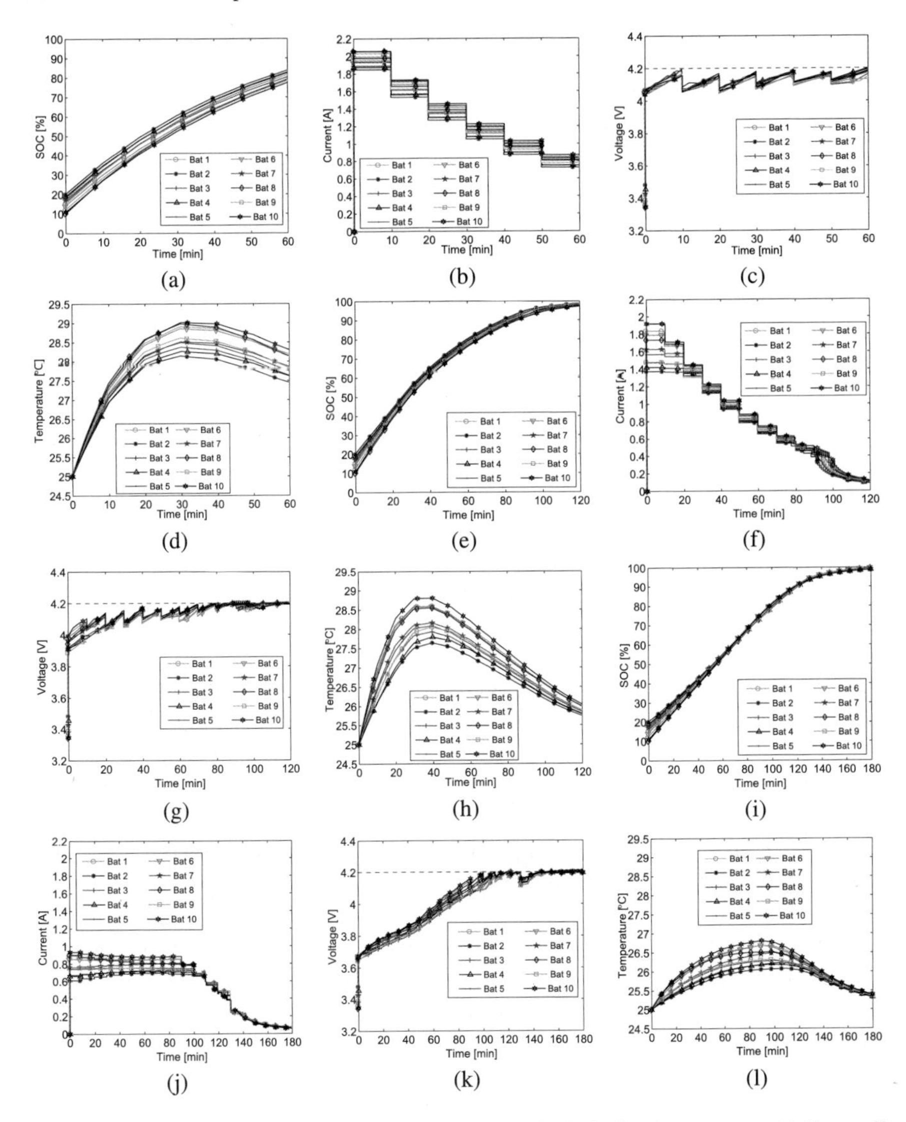

Fig. 9.6 Simulation results of cells' **a** SOCs with $T_{set} = 60$ min, **b** charging currents with $T_{set} = 60$ min, **c** terminal voltages with $T_{set} = 60$ min, **d** temperatures with $T_{set} = 60$ min, **e** SOCs with $T_{set} = 120$ min, **f** charging currents with $T_{set} = 120$ min, **g** terminal voltages with $T_{set} = 120$ min, **h** temperatures with $T_{set} = 120$ min, **i** SOCs with $T_{set} = 180$ min, **j** charging currents with $T_{set} = 180$ min, **k** terminal voltages with $T_{set} = 180$ min, **l** temperatures with $T_{set} = 180$ min

Table 9.1 Simulation results for different user settings

	Average SOC (%)	SOC difference (%)	Maximum temperature (°C)
Initial	15.2	10.83	–
60 min charging	80.5	6.67	29.01
120 min charging	98.01	1.85	28.8
180 min charging	99.36	1.28	26.81

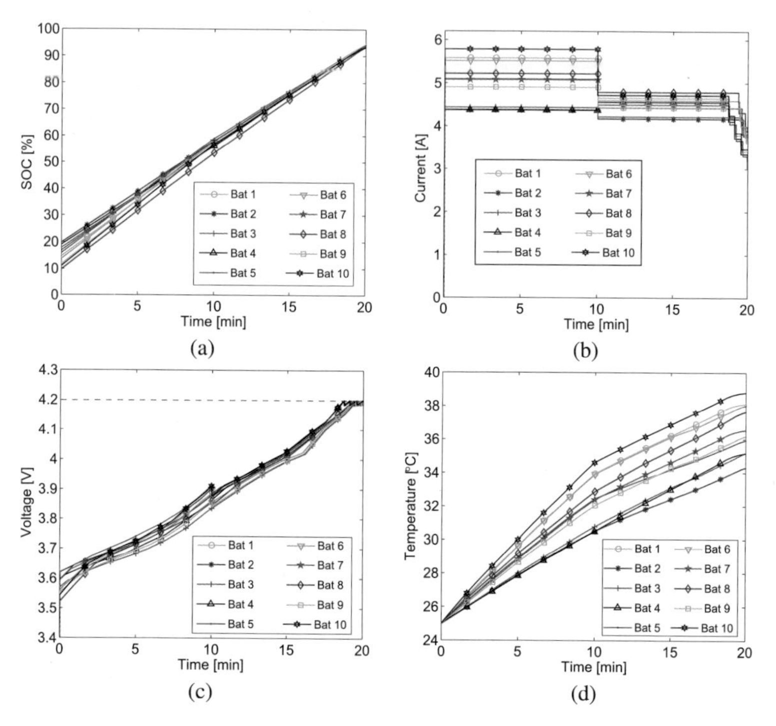

Fig. 9.7 Simulation results of cells' **a** SOCs with $T_{set} = 20$ min, **b** charging currents with $T_{set} = 20$ min, **c** terminal voltages with $T_{set} = 20$ min, **d** temperatures with $T_{set} = 20$ min

show that in our proposed charging approach, fast charging with high C-rate current can be performed while satisfying all charging constraints by setting a short charging duration.

Effect of weights: The weight coefficients in the multi-objective cost function indicate the relative significance of each objective. A large $\gamma_1/\gamma_2/\gamma_3$ increases the importance of the charging task/cell equalization/temperature consideration. The charging task to fulfill user requirements is the most important in the battery charging procedure, indicating that a large γ_1 is required. Different weights of γ_2 and γ_3 are used in the simulation to evaluate the effects of weights on charging performance. To begin, $\gamma_1 = 1000$ and $\gamma_3 = 0.01$ are selected, and γ_2 is chosen as 0.05, 0.1,

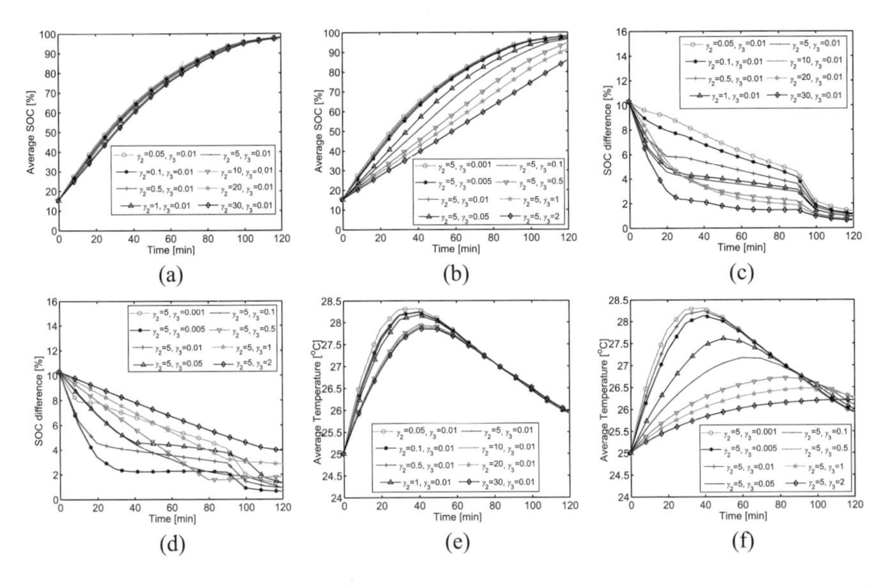

Fig. 9.8 Simulation results of cells' **a** average SOCs with different γ_2, **b** average SOCs with different γ_3, **c** SOC differences with different γ_2, **d** SOC differences with different γ_3, **e** average temperatures with different γ_2, **f** average temperatures with different γ_3

0.5, 1, 5, 10, 20, and 30. The simulation results in terms of average SOCs, SOC differences, and average temperatures are shown in Fig. 9.8a, c, and e. It can be seen that with a large γ_2, less SOC difference can be obtained, which agrees with the above analysis. However, too much γ_2 may degrade the charging task's performance. Then, $\gamma_1 = 1000$ and $\gamma_2 = 5$ are chosen, followed by a new set of γ_3 as 0.001, 0.005, 0.01, 0.05, 0.1, 0.5, 1, and 2, respectively. As shown in Fig. 9.8b, d, and f, a large γ_3 can result in lower average temperature of the cells, but may result in insufficient charging of the battery pack. Overall, the results demonstrate that a high γ_2/γ_3 ratio can result in less cell SOC difference/temperature but may have an impact on the charging task's performance to charge the cells' SOCs to their desired value. Users can choose appropriate weights based on this general rule and use simulation performance as a guide for weight selection in order to balance these objectives in practice.

9.3.2 Experimental Results

Several experiments are carried out on the aforementioned three-modular battery pack to validate the effectiveness of the proposed modified isolated buck converter based multi-module battery charger and the hierarchical control based optimal multi-objective charging strategy. The initial SOCs of the battery cells are set at random to [5%, 15%, 10%] by discharging them from fully charged states. The maximum allowed charging current for the cells is set to 0.5 C, as recommended by the manufacturer.

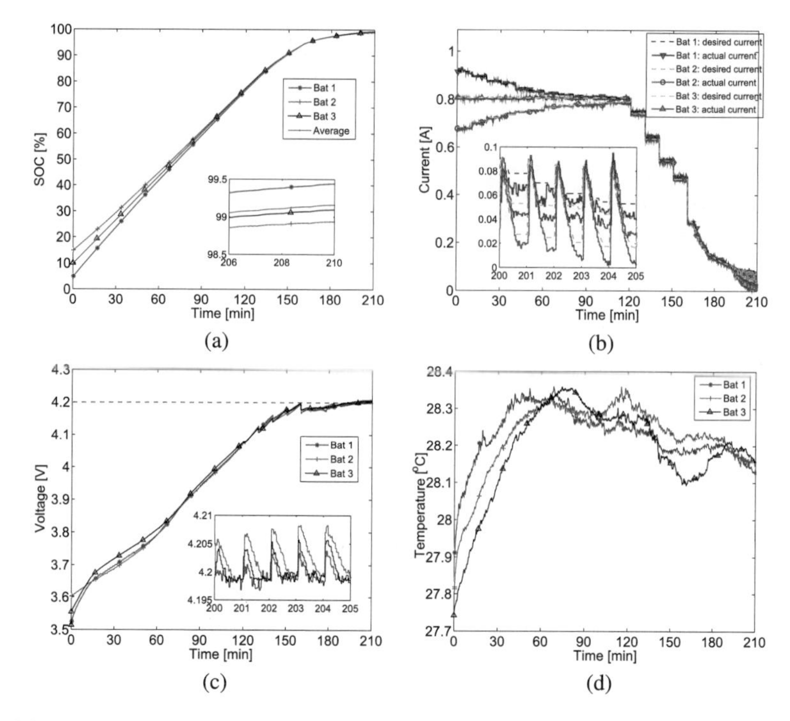

Fig. 9.9 Experimental results of **a** cells' SOC responses, **b** charging currents, **c** terminal voltages, **d** temperatures

The desired SOC of the battery pack is set to $\Gamma_{set} = 100\%$, and the charging duration is set to $T_{set} = 210$ min. The proposed charging current scheduling algorithm is run in about two seconds using a GPIC board.

When a simple internal resistance-based battery model is used in the optimal charging current scheduling algorithm, this might introduce bias when compared to the actual dynamics of the cells. Because of the model bias, the cells' terminal voltages easily surpass their constraints (9.16) with the scheduled charging currents when the cells are charged to near-full charge, as shown in Fig. 9.9a and c. However, by reducing the charging currents by a specific rate $\alpha = 5\%$ each time the cells' measured voltages are greater than the voltage bound, the cells' terminal voltages can be quickly brought close to 4.2 V, demonstrating the benefit of the on-line charging current adjusting approach. The desired charging currents and actual charging current responses of multi-module chargers are depicted in Fig. 9.9b. It demonstrates that the proposed IDA-PBC can be effectively utilized to regulate actual charging currents via the charger in order to track their desired values. Figure 9.9a depicts the cells' SOC responses, with an average SOC of 99.16% at the end of the charging process ($SOC_1 = 99.45$, $SOC_2 = 98.94$, $SOC_3 = 99.09$), which is close to the desired value. The SOC difference between cells decreases from 7.07% to 0.37%. It illustrates that the optimal multi-objective charging approach utilizing hierarchical control can satisfy the charging objectives of user demand and cell equalization. In conventional charging techniques without cell equalization, all cells in the battery pack receive the same amount of charging current. If used for this pack, they would force the charging process to stop when the second cell is fully charged ($SOC_2 = 100\%$) to avoid overcharging, despite the fact that the SOCs of the first and third cells are only $SOC_1 = 91.67\%$ and $SOC_3 = 95.29\%$ (average SOC is 95.65% and SOC difference is 5.91%). It demonstrates the benefit of our designed charging technique with cell equalization over these conventional approach in terms of increasing the effective capacity of the battery pack. The temperatures of the cells are measured by OMEGA temperature sensors, which are kept within a suitable range of [27.7 ℃, 28.4 ℃] during the charging process, as shown in Fig. 9.9d.

References

1. A. Fotouhi, D.J. Auger, K. Propp, S. Longo, M. Wild, A review on electric vehicle battery modelling: from Lithium-ion toward Lithium-Sulphur. Renew. Sustain. Energy Rev. **56**, 1008–1021 (2016)
2. X. Lin, A.G. Stefanopoulou, Y. Li, R.D. Anderson, State of charge imbalance estimation for battery strings under reduced voltage sensing. IEEE Trans. Control Syst. Technol. **23**(3), 1052–1062 (2015)
3. Y. Wang, H. Fang, L. Zhou, T. Wada, Revisiting the state-of-charge estimation for Lithium-ion batteries: a methodical investigation of the extended Kalman filter approach. IEEE Control Syst. Mag. **37**(4), 73–96 (2017)
4. J. Chen, Q. Ouyang, C. Xu, H. Su, Neural network-based state of charge observer design for Lithium-ion batteries. IEEE Trans. Control Syst. Technol. **26**(1), 313–320 (2017)
5. J. Zeng, Z. Zhang, W. Qiao, An interconnection and damping assignment passivity-based controller for a DC-DC boost converter with a constant power load. IEEE Trans. Ind. Appl. **50**(4), 2314–2322 (2014)
6. H. Fang, Y. Wang, J. Chen, Health-aware and user-involved battery charging management for electric vehicles: linear quadratic strategies. IEEE Trans. Control Syst. Technol. **25**(3), 911–923 (2017)
7. A. Abdollahi, X. Han, G. Avvari, N. Raghunathan, B. Balasingam, K.R. Pattipati, Y. Bar-Shalom, Optimal battery charging, part i: minimizing time-to-charge, energy loss, and temperature rise for OCV-resistance battery model. J. Power Sources **303**, 388–398 (2016)
8. S. Boyd, S.P. Boyd, L. Vandenberghe, *Convex Optimization* (Cambridge university Press, 2004)
9. H.K. Khalil, *Nonlinear Systems* (Prentice Hall, Englewood Cliffs, NJ, USA, 2002)

Chapter 10
Simultaneous Charging Equalization Strategy for Battery Packs

10.1 Charging Model

In Fig. 10.1, a generalized diagram of simultaneous charging for the lithium-ion battery packs is provided. Usually, the AC microgrid and some renewable energy resources such as the ocean energy source and the solar energy source are utilized as the power supply to converters. Besides, the converters are composed of n modified CVCS, and each of them is utilized for charging a battery pack. The constant voltage current sources have the advantages of being easy to implement, practical size, and floor price.

10.1.1 Battery Pack Modeling

In this work [1], an internal resistance equivalent circuit model [2, 3] is utilized to shape the characteristic of the battery systems, which achieves the balance between the model accuracy and the computational complexity. The detailed model of battery packs can be found in [4–7]. In order to simplify the design and analyze the experimental results, the influence of battery temperature on the design parameters may be ignored. A controlled voltage source and a connected internal resistance simulate the dynamic characteristic of the single battery, which is presented in Fig. 10.1. According to Fig. 10.1, the terminal voltage of the i-th pack is obtained as

$$V_{B_i}(k) = V_{OC_i}(k) - R_0 I_{B_i}(k) \tag{10.1}$$

where $V_{B_i}(k)$ and $Voc_i(k)$ denote the terminal voltage and the open circuit voltage (OCV) of the i-th battery, respectively. The internal resistances of the battery are assumed to be approximately the same. Considering the manufacturing standards for the same manufacturer, the resistances of battery packs should be more or less constant with some minor deviations. However, in the algorithm a larger resistor

J. Chen et al., *Equalization Control for Lithium-Ion Batteries*,
https://doi.org/10.1007/978-981-99-0220-0_10

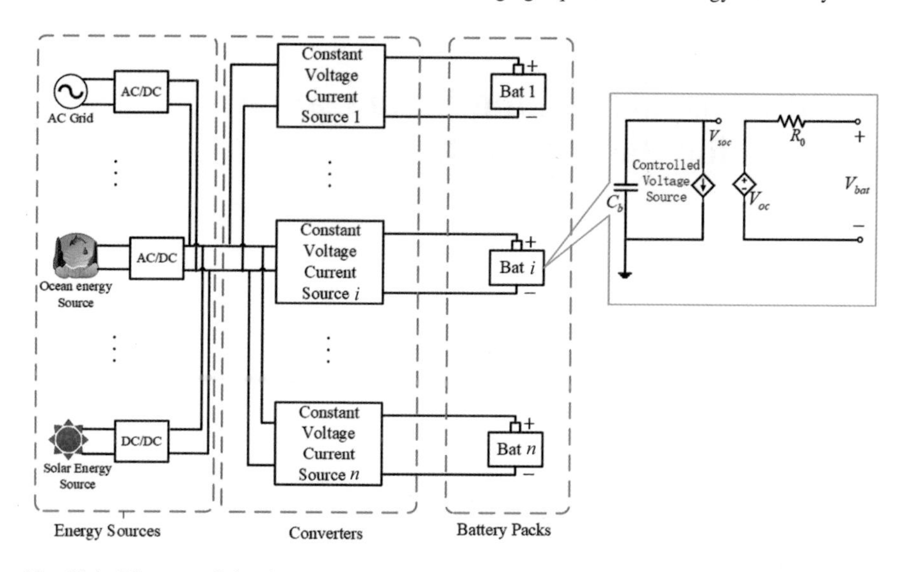

Fig. 10.1 Diagram of simultaneous charging for lithium-ion battery packs

value has been selected than the actual value. The current $I_{B_i}(k)$ denotes the charging current of the i-th battery, and it is defined positive when the battery is in a discharging mode, $Vsoc_i(k)$ is the voltage of the i-th controlled voltage source at time index k, which represents the value of the i-th battery SOC, and $Vsoc_i(k) \in [\,0\text{ V}\,,1\text{ V}\,]$ corresponds to $0 - 100\%$ for the SOC. Usually, there is a nonlinear relationship between the OCV and the SOC of the battery. This relationship could be formulated as a nonlinear function $Voc_i(k) = f(V_{SOC_i}(k))$ with curvefitting, and can be measured when the circuit is open.

The external current of the battery pack from the charger is assumed to be constant during each sampling interval. Referring to [8], the change of i-th $(1 \le i \le n)$ battery's SOC in a sampling period can be calculated as

$$Vsoc_i(k+1) = V_{SOC_i}(k) - \frac{\eta T I_{B_i}(k)}{Q} \tag{10.2}$$

where $Vsoc_i(k+1)$ is the i-th battery's SOC at time index $k+1$, T is the control sampling time period which is much larger than the switching period T_s, η is the Coulombic efficiency, and Q denotes the capacity of the i-th battery. The i-th $(1 \le i \le n)$ battery's SOC is calculated with (10.2). Details on the estimation of SOC can be found in [9–11].

For the i-th $(1 \le i \le n)$ battery pack, the current can be obtained as follows:

$$I_{B_i}(k) = I_i(k) + d_i(k) \tag{10.3}$$

where $d_i(k)$ is the disturbance current of the i-th battery, which is time varying and known, $I_i(k)$ denotes the charging current of the i-th battery, and it is obtained by controlling the output current of the CVCS.

Then, the model of the simultaneous charging system for n battery packs can be represented as a multi-battery model of n identical discrete-time agents as:

$$x_i(k+1) = x_i(k) - \frac{\eta T(u_i(k)+d_i(k))}{Q} \tag{10.4}$$

where $x_i(k) \triangleq V_{SOC_i}(k)$ denotes the system state of the i-th $(1 \le i \le n-1)$ battery, and $u_i(k) = I_i(k)$ is the control input.

The aim of battery packs charging is to design a controller so that the SOCs of batteries converge to the same value. With this control goal, the charging process is defined as follows:

$$\lim_{k \to \infty} \|x_i(k) - x_d\| = 0, \quad \text{for all } i = 1, \ldots, n \tag{10.5}$$

where x_d is a scalar which represents the desired value of x_i in the charging process, and $\|\cdot\|$ is the vector 2-norm. For a dynamic system $x(k+1) = f(x(k), u(k))$, where $u(k)$ is the input of the dynamic system. Moreover, $u(k)$ is said to be the optimal charging process input if charging error $\|x(k) - x_d \times 1_{n\times1}\|$ is minimized at time index k, where $1_{n\times1} \in \mathbb{R}^{n\times1}$ with all entries equal to one. The convergence process can be described as

$$\|x(k) - x_d \times 1_{n\times1}\| > \|x(k+1) - x_d \times 1_{n\times1}\|. \tag{10.6}$$

The modified CVCS can be controlled to achieve simultaneous charging by incorporating the proposed algorithm. The proposed strategy can solve the battery charging issue of large scale battery packs efficiently.

10.1.2 Charging Objective

In the present MOOP formulation of simultaneous charging, four objectives are defined to achieve with the proposed algorithm as listed below:

- The sum of SOC deviation of batteries converges to 0 with battery charging constraints.
- Energy loss of the internal resistance is minimized in the charging mode.
- The batteries' SOCs converge to the same desired value in the charging mode.
- The simultaneous charging time is minimized in the charging mode.

where the third and the fourth objectives are newly designed for battery charging, which is different from [12]. With detailed mathematical derivation, the multi-objective optimization problem can be formulated as [13]:

$$
\begin{aligned}
\min_{x} \quad & F(x) = \left[f_1(x)\ f_2(x)\ f_3(x)\ f_4(x)\right]^T \\
f_1(x) &= \sum_{k=1}^{m}\left(\sum_{i=1}^{n-1}(x_i(k) - x_{i+1}(k))^2 + (x_n(k) - x_1(k))^2\right) \\
f_2(x) &= \sum_{k=1}^{m}\sum_{i=1}^{n} R_0(u_i(k) + d_i(k))^2 \\
f_3(x) &= \sum_{k=1}^{m}\sum_{i=1}^{n}(x_i(k) - x_d)^2 \\
f_4(x) &= \max\{T_1(\varepsilon_1), T_2(\varepsilon_2)\}
\end{aligned}
\tag{10.7}
$$

where $F(x)$ is the vector objective function, and $f_i(x)$ represents the objective function, $i = 1, 2, 3, 4$. The background of this strategy is to solve the problem of non-connected simultaneous charging, the first item of the cost function is set as the difference of the state of charge of each battery pack to ensure simultaneity. The second term is set as energy loss to limit excessive charging current. The third function is set as the difference between the states of charge of each battery pack and the expected states to achieve the goal of being fully charged at the same time. The fourth term is set as the charging time, because it is expected to shorten the total charging time while satisfying simultaneous charging. Function $f_1(x)$ represents the deviation sum of the SOC among two adjacent batteries, $f_2(x)$ defines the Ohmic loss on the series resistance of the batteries, and $f_3(x)$ is the error between the actual SOC of the batteries and the target value of SOC, which is different from the previous work [12]. In [12], we only consider the difference of SOCs and the power consumption. However, in this work, the third objective is taken into consideration while designing the charging strategy as a more practical objective. Function $f_4(x)$ represents the simultaneous charging time. In (10.7), $T_1(\varepsilon_1)$ and $T_2(\varepsilon_2)$ are the converging time and the charging time under a certain error, and are defined as

$$
\begin{aligned}
T_1(\varepsilon_1) &= \min\{\tau |\ \|x_i(k) - x_j(k)\| \leqslant \varepsilon_1, \forall k > \tau/T, \forall i, j\} \\
T_2(\varepsilon_2) &= \min\{\tau |\ \|x(k) - x_d\| \leqslant \varepsilon_2, \forall k > \tau/T\}.
\end{aligned}
\tag{10.8}
$$

The simultaneous charging time T is defined as $T = \max\{T_1(\varepsilon_1), T_2(\varepsilon_2)\}$ in a simultaneous charging and converging process.

10.1.3 Charging Constraints

The following three constraints should be satisfied in the charging process to guarantee the stability of the battery pack system and extend battery lifetime: the SOC limitation of batteries, the charging current constraints, and the terminal voltage constraints. To avoid overcharging of the battery packs, the batteries' SOCs should be limited within a certain range as follows:

$$\begin{aligned} x_i(k+1) =& x_i(k) - su_i(k) - \theta_i(k) \\ x_i^l(k) \leqslant& x_i(k) \leqslant x_i^u(k) \end{aligned} \tag{10.9}$$

where $s \triangleq \eta T/Q$, $\theta_i(k) \triangleq \eta T d_i(k)/Q$, $x_i^l(k)$ and $x_i^u(k)$ represent the lower bound and the upper bound of the i-th battery SOC at time index k. The excessive charging current is harmful to the batteries and can be defined as the charging current over the maximum allowable limit. Therefore, the charging current of batteries should be maintained within a suitable range that

$$I_i^l(k) \leqslant u_i(k) + d_i(k) \leqslant I_i^u(k) \tag{10.10}$$

where $I_i^l(k)$ and $I_i^u(k)$ represent the minimum and the maximum of $I_i(k)$, which is constrained by the actual circuit. In particular, considering the maximum power limit of the charging station, $I_i^u(k)$ can be defined as

$$I_i^u(k) = \min\left(I_{max}, I_i(k-1) + \Delta I_{max}, \frac{P_{max}}{n \cdot V_{B,min(k)}}\right) \tag{10.11}$$

where P_{max} is the maximum power limitation. The variables I_{max} and ΔI_{max} represent the maximum current limit of charging and the maximum rate of change of current per second, respectively. Moreover, the terminal voltage of batteries can not exceed the maximum voltage during charging, due to the availability of voltage protection mechanisms. Therefore, (10.12) can be obtained as follows:

$$V_{OC_i}(k) + R_0 I_i(k) \leqslant V_{max} \tag{10.12}$$

where V_{max} represents the rated voltage of the batteries.

10.2 Simultaneous Charging Development

This section will explain the optimal charging strategy of the battery packs. For a nontrivial multi-objective optimization problem, there is no single solution that can optimize each objective simultaneously. A multi-objective vector valued function consists of two or more conflicting objectives, and hence the optimization of MOOP generates a Pareto front with non-dominated optimal solutions by making trade-offs between objective functions [13, 14]. In order to solve the multi-objective optimization problem, some of the most used methods, such as the scalarizing method, no-preference method, and interactive methods, are utilized commonly in the literature. In this paper, linear scalarization is incorporated to solve the multi-objective optimization problem. In general, the linear scalarization method is incorporated to assign the proper weightage to the objective functions according to the needs for scalarizing the multiple performance objectives to obtain a single solution. Considering that the simultaneous charging time can not be determined until the charging process will be over. In this case, the multi-objective optimization problem can be rewritten as a hierarchical multi-objective optimization problem. The detailed algorithm for solving the multi-objective optimization is divided into two processes:

- After linear scalarization of the multi-objective optimization problem, a set of weight coefficients are generated. Then, the control input $u_{i,k}$ is obtained by minimizing the weighted sum of the first three objectives $f_1(x)$, $f_2(x)$, and $f_3(x)$, a balancing or charging time cost function can be obtained.
- Based on the one-dimensional optimization of weights, minimum of the simultaneous charging time can be obtained. Suppose that there are n batteries, and every battery has an initial SOC $x_i(0), i = 1, 2, ..., n$. Considering the ideology of the predictive control [15], the optimal control inputs over m steps are calculated by solving (10.7), and only the first step optimal prediction method is then employed as the control input.

With these two assumptions, the optimization problem can be formulated as:

$$\begin{aligned}(T_1, T_2) = &\arg\min_{u_{i,j}\in\Omega} \sum_{j=1}^{m}\Bigg(\sum_{i=1}^{n-1}(x_i(j) - x_{i+1}(j))^2 + (x_n(j) - x_1(j))^2 \\ &+ \alpha\sum_{i=1}^{n}(x_i(j) - x_d)^2 + \beta\sum_{i=1}^{n} R_0(u_i(j-1) + d_i(j))^2\Bigg) \\ (\alpha, \beta) = &\arg\min_{\alpha,\beta}\max\{T_1, T_2\} \\ x_i(j+1) = &x_i(j) - su_i(j) - \theta_i(j), i = 1, \ldots, n, \ j = 1, \ldots, m\end{aligned} \tag{10.13}$$

where α and β are two positive weight coefficients. The first objective has a quadratic form, which can be rewritten as quadratic programming as follows:

$$\begin{aligned}\min\ J = &\sum_{j=1}^{m}\Bigg[\sum_{i=1}^{n-1}(x_i(j) - x_{i+1}(j))^2 + (x_n(j) - x_1(j))^2 \\ &+\alpha\sum_{i=1}^{n}(x_i(j) - x_d)^2 + \beta\sum_{i=1}^{n} R_0(u_i(j-1) + d_j)^2\Bigg] \\ = &\sum_{j=1}^{m}\Big(\frac{1}{2}X_j^T P X_j + q^T X_j + \frac{1}{2}U_{j-1}^T Q U_{j-1} + r_j^T U_{j-1} \\ &+ n\alpha x_d^2 + n\beta R_0 d_j^2\Big) \\ = &\frac{1}{2}\mathscr{X}^T H \mathscr{X} + f^T\mathscr{X} + mn\alpha x_d^2 + mn\beta R_0 d_j^2\end{aligned} \tag{10.14}$$

where the intermediate variable X_j is defined as $X_j = [\,x_1(j), \ldots, x_n(j)]^T \in \mathbb{R}^n$, which represents the SOCs of batteries at j-th time. The intermediate control variable

U_j is defined as $U_j = [u_1(j), \ldots, u_n(j)]^T \in \mathbb{R}^n$, which is the control input of batteries at j-th control period. The quadratic term coefficient matrix of the intermediate variable is given as follows:

$$P \triangleq \begin{bmatrix} 4+2\alpha & -2 & 0 & \cdots & -2 \\ -2 & 4+2\alpha & -2 & \cdots & 0 \\ 0 & -2 & 4+2\alpha & \cdots & 0 \\ \vdots & \vdots & \vdots & \ddots & \vdots \\ -2 & 0 & \ldots & 0 & 4+2\alpha \end{bmatrix}_{n \times n}$$

and the variable linear term is defined as $q \triangleq [-2\alpha x_d, \ldots, -2\alpha x_d]^T \in \mathbb{R}^n$. The quadratic term coefficient matrix of the intermediate control variable is defined as $Q = \text{diag}(2\beta R_0, \ldots, 2\beta R_0) \in \mathbb{R}^{n \times n}$, with diag(·) denoting the diagonal matrix. Moreover, the linear term of the intermediate control variable is written as $r_j = [2\beta R_0 d_j, \ldots, 2\beta R_0 d_j]^T \in \mathbb{R}^n$. Then, the variable $\mathscr{X}$ is defined as $\mathscr{X} = [X_1^T, \ldots, X_m^T, U_0^T, \ldots, U_{m-1}^T]^T \in \mathbb{R}^{2mn}$, and the corresponding coefficient matrix H is defined as $H \triangleq \text{diag}(P, \ldots, P, Q, \ldots, Q) \in \mathbb{R}^{2mn \times 2mn}$, eventually, and the linear term is defined as $f \triangleq [q^T, \ldots, q^T, r_1^T, \ldots, r_m^T]^T \in \mathbb{R}^{2mn}$.

The inequality optimization constraints of (10.14) can be derived from (10.9), (10.10), and (10.12), which are described as

$$\begin{aligned} (I - A)\mathscr{X} &= \mathscr{X}_0 - \Theta \\ \mathscr{X}_{min} &\preceq \mathscr{X} \preceq \mathscr{X}_{max} \\ V_{OC} + R_0\mathscr{X} &+ R_0\Theta \preceq V_r \end{aligned} \tag{10.15}$$

where $A \triangleq \text{diag}(s, \ldots, s) \in \mathbb{R}^{2mn \times 2mn}$, $\mathscr{X}_0$ represents the initial value of the control variable $\mathscr{X}$, and Θ represents the load disturbance current vector. The relation $\preceq$ denotes a partial order [16], which means that the corresponding elements satisfy the inequality. The corresponding definition of these variables in (10.15) are listed as follows:

$$\begin{aligned} X_{min} =& [x_1^l(1) \ldots x_n^l(1) \ldots x_1^l(m) \ldots x_n^l(m) \\ & \ldots I_1^l(1) \ldots I_n^l(1) \ldots I_1^l(m) \ldots I_n^l(m)]^T \in \mathbb{R}^{2mn} \\ X_{max} =& [x_1^u(1) \ldots x_n^u(1) \ldots x_1^u(m) \ldots x_n^u(m) \\ & \ldots I_1^u(1) \ldots I_n^u(1) \ldots I_1^u(m) \ldots I_n^u(m)]^T \in \mathbb{R}^{2mn} \\ V_{OC} =& [\underbrace{0 \ldots 0}_{mn}, \underbrace{V_{OC} \ldots V_{OC}}_{mn}]^T \in \mathbb{R}^{2mn} \\ V_r =& [\underbrace{0 \ldots 0}_{mn}, \underbrace{V_r \ldots V_r}_{mn}]^T \in \mathbb{R}^{2mn} \\ \Theta =& [\underbrace{0 \ldots 0}_{mn}, \underbrace{d_1(1) \ldots d_n(1) \ldots d_1(m) \ldots d_n(m)}_{mn}]^T \in \mathbb{R}^{2mn}. \end{aligned} \tag{10.16}$$

Therefore, the optimization problem (10.14) can be described as a quadratic programming, which can be solved by the active set method or interior point method [17, 18].

Then, a one-dimensional optimization algorithm for optimizing the simultaneous charging time is introduced. A modified steepest descent algorithm with momentum is proposed to solve the second optimization problem in (10.13). Besides, to accelerate the convergence speed, an adaptive learning rate is developed in the algorithm. Then, an adaptive momentum-based steepest descent algorithm for the second optimization problem in (10.13) is designed as Algorithm 1.

Algorithm 1:

Require:

initial value $\alpha(0) > 0$, $\beta(0) > 0$, $\varepsilon > 0$, $\varepsilon_1 > 0$, $\varepsilon_2 > 0$, $T(0) = +\infty$;

Ensure:

$|T(k) - T(k-1)| \leqslant \varepsilon$ is satisfied;

1: $T_1(\varepsilon_1)$ and $T_2(\varepsilon_2)$ are obtained from (10.14);

2: $T(k) = \max\{T_1, T_2\}$;

3: $\Delta\alpha(k) = -\,\omega(k)(1-\theta)\nabla T(k) + \theta\Delta\alpha(k-1)$;

4: $\Delta\beta(k) = -\,\omega(k)(1-\theta)\nabla T(k) + \theta\Delta\beta(k-1)$;

5: $\omega(k+1) = \begin{cases} \lambda\omega(k), & \nabla T(k) \geqslant \mu\nabla T(k-1) \\ \frac{1}{\lambda}\omega(k), & \nabla T(k) \leqslant \frac{1}{\mu}\nabla T(k-1) \\ \omega(k), & \text{others.} \end{cases}$

In Algorithm 1, an adaptive momentum-based steepest descent algorithm is proposed to optimize the weights α and β, where θ is a momentum factor, $\omega(k)$ is an adaptive learning rate. Compared with the normal steepest descent algorithm, the steepest descent algorithm with momentum has one more term $\theta\Delta\alpha(k-1)$, which means the update direction and amplitude of the weights are not only related to the gradient obtained in this calculation, but also related to the direction and amplitude of the last updating. The iteration is also done according to the historical term $\theta\Delta\alpha(k-1)$ so that the update becomes larger, which makes faster iteration and faster convergence. In the third and fourth steps of Algorithm 1, $\nabla T(k)$ is the gradient of simultaneous charging time, as k increases, $\nabla T(k)$ becomes smaller and smaller. The iteration holds the farther updates by a small dimension and recent updates by a larger dimension.

The addition of the adaptive learning rate $\omega(k)$ makes updation of the weight having certain inertia, as well as certain robustness and accelerating convergence ability. Besides, a variable learning rate is used to accelerate the convergence speed. The idea is that the corresponding parameters receiving large updates will reduce the effective learning rate $\omega(k)$, while the parameters receiving small updates will improve its effective learning rate. In this way, we can accelerate convergence by accelerating the learning rate of both weights α and β. The proposed strategy is an improvement upon the steepest descent by incorporation of both the momentum and the adaptive learning to the update rule.

When the converging time and charging time approach the target value in a reduced way, it indicates that the correction direction is on the right track, and the learning rate can be increased. When the error increases beyond a certain range, it indicates that the previous step correction was not right, the step size should be reduced, and the previous step correction process shall be canceled.

The optimization problem (10.13) of weight coefficients is not convex. The traditional gradient descent method is difficult to guarantee that the final result will be a global minimum. The solution process is straightforward to make convergence at the local minima, and therefore, the computing time becomes longer. The steepest descent method with momentum can continue calculating even though the gradient approaches zero. However, the steepest descent method with momentum is a weighted accumulation, and it will not jump out when the field near the solution is too steep. It is generally considered that the gradient descent method with momentum is suitable to solve the relatively smooth optimization problem.

10.3 Simulation and Experimental Results

In this section, the proposed strategy of simultaneous charging is validated through extensive simulations and experiments.

10.3.1 Simulation Results

In order to verify the effectiveness of the above proposed strategy, a PLECS simulation platform is used for four-modular battery packs with four modified constant voltage current sources (CVCS), where PLECS facilitates the modeling and simulation of complete systems, including power sources, power converters, and loads, which can be integrated with Simulink. The parameters selected in the simulation are illustrated in Table 10.1. Model parameters of the batteries are chosen as [6].

Figure 10.3a–b show the dynamic responses of SOCs and the input control current in a discharging process, where weight coefficients in case of Fig. 10.3a,Fig. 10.3c, and Fig. 10.3e are $\{\alpha = 1, \beta = 10^{-5}\}$, $\{\alpha = 1, \beta = 10^{-4}\}$, and $\{\alpha = 2, \beta = 10^{-6}\}$, respectively. Compared with quasi-sliding mode control [19], which is our previous work, equilibrium time reduces by more than half. The corresponding curves of charging speed and charging time are shown in Fig. 10.3a–f under the influence of different parameters and different terminal constraints. At the beginning of the charging period, all batteries are charged with the maximal currents, with the current constraints being satisfied. With the increase of the SOC, the open circuit voltage of the batteries is also increased. Therefore, the range of the charging current is also becoming smaller and smaller. Figure 10.4 shows the charging error of different

Table 10.1 Selected parameters in the simulation

Definition	Value	Unit
Capacity of the batteries	3.1	Ah
Voltage of the CVCS	5	V
PWM period	1/7000	s
Control period	1	s
Coulombic efficiency	0.9	–

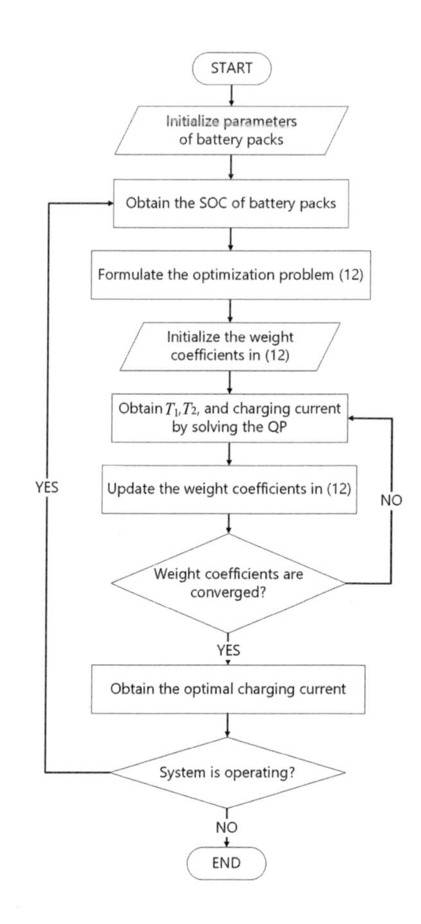

Fig. 10.2 The flow chart of the proposed simultaneous charging strategy for battery packs

batteries with simultaneous charging. It can be seen that with the increase of the charging weight coefficient, charging error converges to 0 at a faster rate. When the charging weight coefficient becomes greater than 5, then by increasing the charging coefficient has no significant effect on the error decrement.

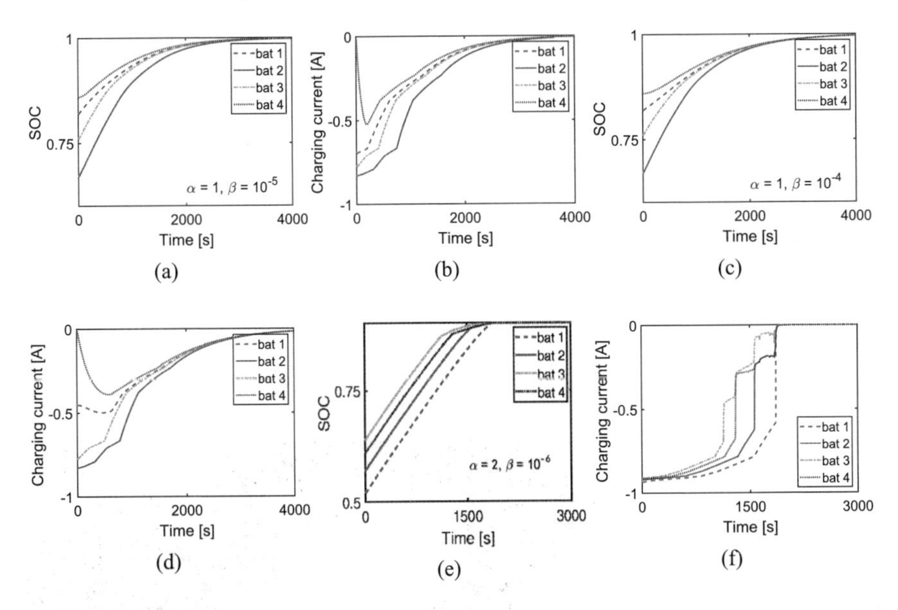

Fig. 10.3 SOC response and input control current response with different weight coefficients and different charging terminal constraints

Fig. 10.4 Charging error for different charging coefficients

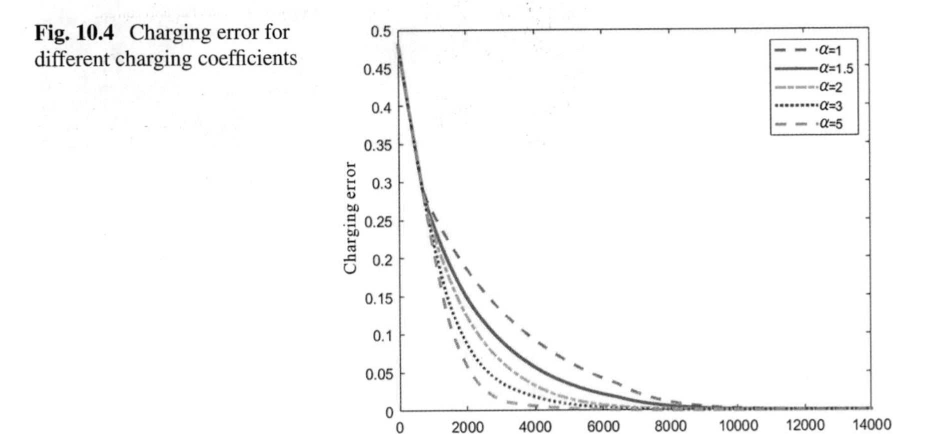

10.3.2 Experimental Results

The simultaneous charging strategy is more engineering oriented, so a detailed experimental verification is carried out to show that the charging strategy is much more practical and more implementable than the conference paper without experiments [12]. Lithium-ion battery packs composed of four NCR 18650 (MH12210–3100 mAh) lithium-ion batteries are utilized in the experimental platform, as illustrated

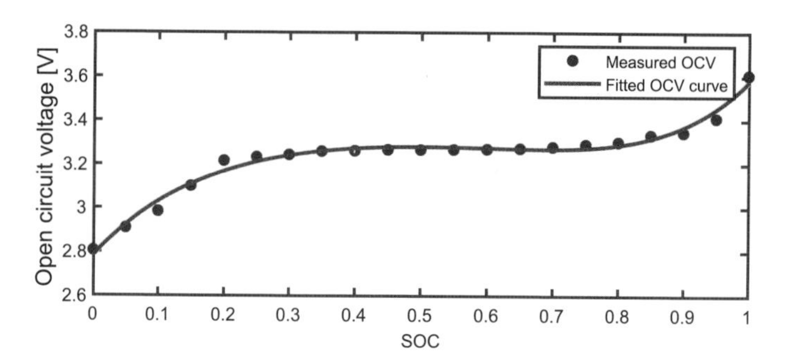

Fig. 10.5 Relationship between the open circuit voltage and the SOC

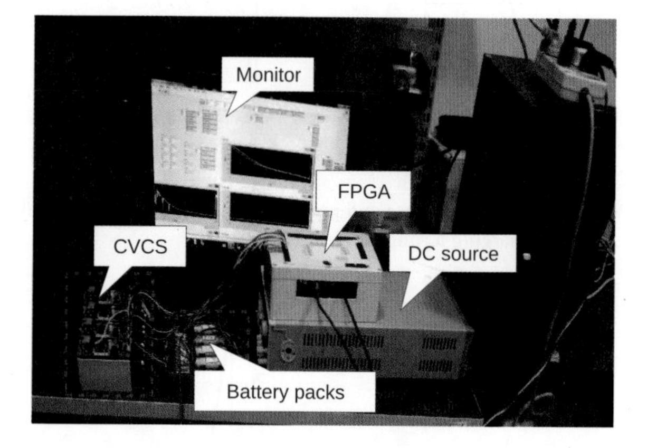

Fig. 10.6 The experimental platform of the battery simultaneous charging system

in Fig. 10.6. In order to simplify the experiment setup and reduce the difficulty of the experiment, the batteries are treated as battery packs. The upper part of Fig. 10.6 is the power supply module and the measurement module, and the lower part is batteries and constant voltage current sources for the batteries. Through several cycles of charging test, the capacities of these batteries are identified as 3.1 Ah and $Q = 11160$ C. The relationship between the open circuit voltage and the SOC of these batteries can be seen in Fig. 10.5 (Note, it is better to change the sequence of Figs. 10.5 and 10.6).

Then, two different cycles are carried out in the experiments. Figure 10.2 is the flow chart of the proposed charging strategy for the battery packs. Figures 10.7 and 10.8 show the experimental results of the simultaneous charging of four battery packs using the fixed weight coefficients and proposed optimization charging strategy, respectively. In Fig. 10.7a, weight coefficients α and β are set to $\alpha = 2, \beta = 10^{-4}$. In this case, the simultaneous full charging time is close to 10000 s, and the converging time is 9562 s. The relative error of the time is near to 5%. Moreover, with the utilization of the adaptive momentum-based steepest descent algorithm, the simultaneous charging time and the converging time are 5583 s and 5533 s, as shown

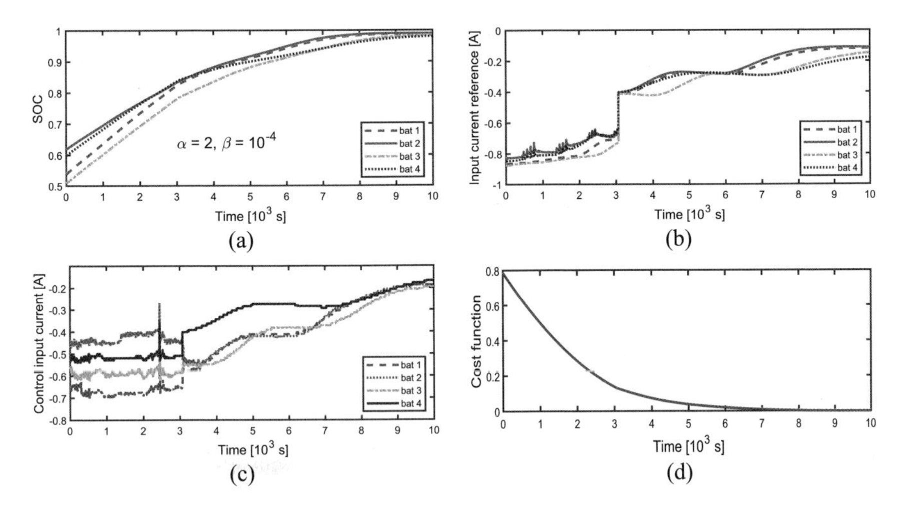

Fig. 10.7 When weight coefficients are $\alpha = 2$ and $\beta = 10^{-4}$, **a** SOCs in simultaneous charging mode, **b** input current reference in simultaneous charging mode, **c** control input current in simultaneous charging mode, **d** Cost function in (10.14)

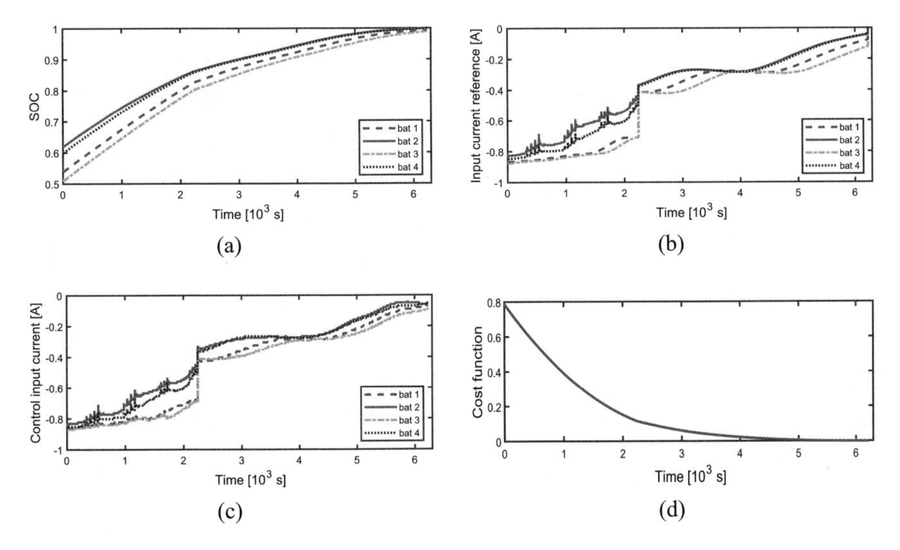

Fig. 10.8 When weight coefficients are updated by the proposed strategy, **a** SOCs in simultaneous charging mode, **b** input current reference in simultaneous charging mode, **c** control input current, **d** Cost function in simultaneous charging mode, with optimal weight coefficients

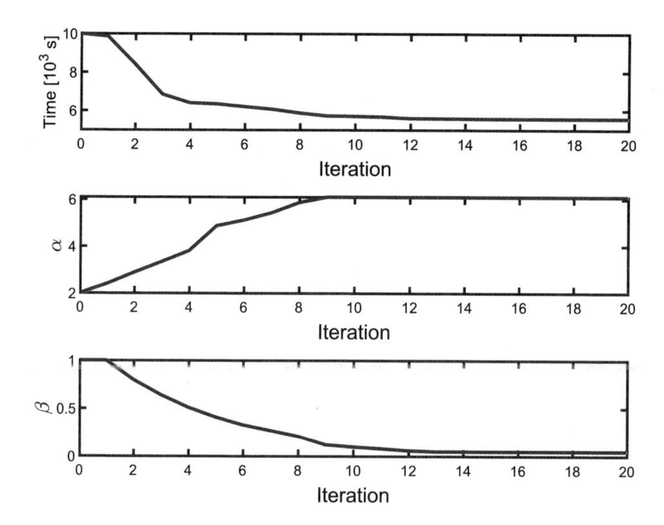

Fig. 10.9 Simultaneous charging time and two weight coefficients with the adaptive momentum based steepest descent algorithm

in Fig. 10.8a–c. Therefore, the charging time is decreased greatly in the charging process, and the relative time error of simultaneous time and converging time has also decreased to less than 1%. In [20, 21], the relative error between the simultaneous time and the converging time is more than 8%. Therefore, as compared to the relative time error, it can be seen that the performance of the proposed charging strategy is much better than the battery-assisted charging system in [20]. The proposed strategy has increased the charging process's efficiency and speed. Moreover, the utilization of the proposed strategy avoids overcharging and overdischarging, which is one of the important factors that cause battery degradation [20]. The strictness of charging and charging current can prevent overcharging more effectively and smoothly in real time feedback under the model predictive control framework. Correspondingly, the proposed strategy can effectively reduce the battery degradation. Furthermore, the charging current is smoother than before, which means that the current is more friendly to the CVCS. Figures 10.7d and 10.8d represent the change of the cost function in (10.14), we can see that the cost function converges to 0 rapidly with the AMSDA. Besides, keeping decreasing the cost function also shows the effectiveness of the proposed charging strategy.

Figures 10.7 and 10.8 show the performance of different weight parameters. Figure 10.9 shows the performance of the proposed Algorithm 1. In Algorithm 1, the adaptive momentum-based steepest descent is incorporated to choose a proper weight parameter for the simultaneous charging. In Fig. 10.9, we can see that the simultaneous charging time decreased rapidly with few iterations. Furthermore, the parameters α and β also converge to the optimal solutions. Then, the optimization problem (10.13) is transformed as a normal quadratic programming problem with

Table 10.2 Computational costs

Number of Battery Packs	Prediction Horizon	CPU Time (s)
10	10	0.0112
30	10	0.0650
50	10	0.2239
100	10	1.3036

the choice of the weight coefficients. In order to verify the real-time performance of the algorithm, it is tested for different sizes of batteries with the same prediction step. It is worth mentioning that the horizontal coordinates of Fig. 10.7d and 10.8d do not refer to the time required for one calculation, but the duration of the entire experiment. The computational costs of this algorithm are shown in the Table 10.2.

References

1. J. Chen, H. Chen, M. Zhou, L. Kumar, J. Zheng, Quadratic programming based simultaneous charging strategy for battery packs of electric vehicles. IEEE/ASME Trans. Mech. **27**(6), 5869–5878 (2022)
2. X.Lin, A.G. Stefanopoulou, Y. Li, R.D. Anderson, State of charge imbalance estimation for battery strings under reduced voltage sensing. IEEE Trans. Control. Syst. Technol. **23**(3), 1052–1062 (2015)
3. X. Cui, W. Shen, Y. Zhang, C. Hu, A novel active online state of charge based balancing approach for lithium-ion battery packs during fast charging process in electric vehicles. Energ. **10**(11), 1766–1773 (2017)
4. Q. Ouyang, J. Chen, J. Zheng, H. Fang, Optimal multiobjective charging for lithium-ion battery packs: a hierarchical control approach. IEEE Trans. Ind. Inform. **14**(9), 4243–4253 (2018)
5. J. Chen, Q. Ouyang, C. Xu, H. Su, Neural network-based state of charge observer design for lithium-ion batteries. IEEE Trans. Control. Syst. Technol. **26**(1), 313–320 (2017)
6. M. Chen, G. Rincon-Mora, Accurate electrical battery model capable of predicting runtime and $I-V$ performance. IEEE Trans. Energy Convers. **21**(2), 504–511 (2006)
7. H. Fang, Y. Wang, J. Chen, Health-aware and user-involved battery charging management for electric vehicles: linear quadratic strategies. IEEE Trans. Control. Syst. Technol. **25**(3), 911–923 (2017)
8. H. He, R. Xiong, X. Zhang, F. Sun, J. Fan, State-of-charge estimation of the lithium-ion battery using an adaptive extended Kalman filter based on an improved thevenin model. IEEE Trans. Control. Syst. Technol. **60**(4), 1461–1469 (2011)
9. D.D. Domenico, E. Prada, Y. Creff, An adaptive strategy for li-ion battery internal state estimation. IEEE Trans. Control. Syst. Technol. **21**(12), 1851–1859 (2013)
10. H. Fang, Y. Wang, Z. Sahinoglub, T. Wada, S. Hara, State of charge estimation for lithium-ion batteries: an adaptive approach. Control. Eng. Pract. **25**, 45–54 (2014)
11. H. Fang, X. Zhao, Y. Wang, Z. Sahinoglu, T. Wada, S. Hara, R.A. de Callafon, Improved adaptive state-of-charge estimation for batteries using a multi-model approach. J. Power Sources **254**, 258–267 (2014)
12. H. Chen, X. Fan, J. Zheng, Y. Fu, J. Chen, Optimal multi-objective cell balancing for battery packs with quadratic programming, in *IEEE 28th International Symposium on Industrial Electronics (ISIE)*, (Vancouver, BC, Canada, 2019), pp. 1989–1994

13. K. Miettinen, *Nonlinear Multiobjective Optimization* (Springer, Berlin, Germany, 2012)
14. H. Chen, J. Chen, H. Lu, C. Yan, Z. Liu, A modified MPC-based optimal strategy of power management for fuel cell hybrid vehicles. IEEE/ASME Trans. Mechatron. **25**(4), 2009–2018 (2020)
15. H. Zheng, R.R. Negenborn, G. Lodewijks, Fast ADMM for distributed model predictive control of cooperative waterborne AGVs. IEEE Trans. Control. Syst. Technol. **25**(4), 1406–1413 (2017)
16. J.J. Rotman, *Advanced Modern Algebra* (American Mathematical Society, Providence, Rhode Island, 2010)
17. P.R. Takács, Z. Darvay, A primal-dual interior-point algorithm for symmetric optimization based on a new method for finding search directions. Optim. **67**(6), 889–905 (2018)
18. H.J. Ferreau, C. Kirches, A. Potschka, H.G. Bock, M. Diehl, qpOASES: a parametric active-set algorithm for quadratic programming. Math. Program. Comput. **6**(4), 327–363 (2014)
19. Q. Ouyang, J. Chen, J. Zheng, Y. Hong, Soc estimation-based quasi-sliding mode control for cell balancing in lithium-ion battery packs. IEEE Trans. Ind. Electron. **65**(4), 3427–3436 (2018)
20. M. Aziz, T. Oda, M. Ito, Battery-assisted charging system for simultaneous charging of electric vehicles. Energy **100**, 82–90 (2016)
21. M. Aziz, T. Oda, Simultaneous quick-charging system for electric vehicle. Energy Procedia. **142**, 1811–1816 (2017)